Tatyane Fernandes
Fabiana Silva Hussein

Teaching the visually impaired from a multi-sensory perspective

AF301533

Tatyane Fernandes
Fabiana Silva Hussein

Teaching the visually impaired from a multi-sensory perspective

The importance of experimentation and computer programmes for more inclusive teaching

ScienciaScripts

Imprint

Any brand names and product names mentioned in this book are subject to trademark, brand or patent protection and are trademarks or registered trademarks of their respective holders. The use of brand names, product names, common names, trade names, product descriptions etc. even without a particular marking in this work is in no way to be construed to mean that such names may be regarded as unrestricted in respect of trademark and brand protection legislation and could thus be used by anyone.

Cover image: www.ingimage.com

This book is a translation from the original published under ISBN 978-613-9-66524-2.

Publisher:
Sciencia Scripts
is a trademark of
Dodo Books Indian Ocean Ltd. and OmniScriptum S.R.L publishing group

120 High Road, East Finchley, London, N2 9ED, United Kingdom
Str. Armeneasca 28/1, office 1, Chisinau MD-2012, Republic of Moldova, Europe
Printed at: see last page
ISBN: 978-620-8-04441-1

"It's not enough to see the other person's difficulty, to be aware of their own condition of inequality, but to have our eyes blinded to our inflexibility of ideas, beliefs and procedures, the lack of social and political meaning of what we teach in the lives of the subjects."

Paulo Ricardo Ross

ACKNOWLEDGEMENTS

I would firstly like to thank God for bringing me peace and tranquillity in difficult times, giving me the strength to concentrate and fight for my goals despite all the obstacles.

I would also like to thank my family for supporting my decisions, supporting me in my difficulties, apologising for my absence at times and being my foundation.

I would also like to thank all the teachers on the FCET postgraduate programme for showing me the best way forward and teaching me to open my mind beyond the world of exact sciences.

My special thanks go to my supervisor Prof Dr Fabiana R.G. e Silva Hussein for her help, support and, above all, her belief in my research.

Finally, I would like to thank the Rio Branco State College for authorising the implementation of the sequence of activities and the research.

SUMMARY

For some years now, the transfer of students with special needs from special education to mainstream education has been taking place in Brazil. However, schools and educators have not been prepared for this change, and physical and attitudinal barriers to inclusion still prevail. It is believed in this research that one of the ways to solve this problem is to work on the inclusion of blind students through experimentation and digital teaching resources. In the scientific literature, there are already several teaching resources that have been created or adapted, but there is still a lot to be done, especially for students with special needs. Considering Vygotsky's theory, in which students with special needs should learn the same content with the same degree of demand as others, and using a multisensory methodology, didactic material was developed consisting of a didactic sequence and adapted materials with the aim of facilitating the teaching-learning process of the chemical reactions content for students with or without sight problems. The research analysing the material developed was qualitative and was carried out with second year secondary school students. The creation of an accessible computer programme and its association with experimentation, as well as the sequence of activities, proved to be efficient in improving learning and the effective inclusion of the visually impaired.

INTRODUCTION

Special education is a subject that has been present in my education from a very early age, as I have always had contact with people with special educational needs. This is because my mother is a teacher at a special school and works with mentally handicapped people.

When I was still at secondary school, I worked as an assistant in the same school with children aged between 1 and 7. In this context, what I observed was that even mentally handicapped pupils have abilities that can be explored and possibilities to attend mainstream school, with the same opportunities as other children.

While I was studying at university, the process of inclusion was put into practice in classrooms all over the country, but as a teacher I realised that this inclusion was only seen as inserting the student into the regular class. This means that the teachers, staff and pedagogical team were not prepared for this, because I believe that in order to meet the needs of special students, a minimum level of training is required, or the student will continue to be excluded, since they don't have the knowledge and tools to promote inclusion.

In the years I've been teaching, I've worked with 4 visually impaired students, 2 hearing impaired and 1 with Down's syndrome, which led me to research and read about the subject in order to prepare lessons suited to their needs. It was then that I realised that in the field of Chemistry there isn't much research into inclusive education in Brazil.

Considering that in chemistry there are many concepts related to the sense of sight, I thought that teaching chemistry to the visually impaired could be a research problem. I believe that exploring this topic could help teachers who are faced with this situation in the classroom to have more support for their practice.

Developing action research has greatly enriched my knowledge of various concepts related to chemistry teaching and inclusive education. I came across some research that I didn't know about, and which helped me a lot not only in carrying out the work presented in this dissertation, but also in my teaching practice. I can say with certainty that my perception has been transformed and that my personal enrichment has gone beyond expectations.

INTERVIEWING PROFa. MIRIA FAGUNDES

In an interview with Prof Miria de Souza Fagundes, from the visual impairment team at SEED's Department of Special Education and Educational Inclusion, we tried to clarify the role of the Department of Education in the inclusion process and what resources have been made available.

The first piece of information requested was the number of DV students enrolled in Paraná's state school system. According to Miria, there are 698 blind students and 5700 low vision students. So the study and development of methodologies to work with this public is important, because the number of students is not as small as some people imagine.

Brazilian legislation stipulates that pupils with special educational needs have the right to quality education, preferably in mainstream schools. However, the interviewee explains that this doesn't always happen, because sometimes children are left without access to school, due to a lack of school transport, specialised teachers, services and support, or their parents, in an overprotective attitude, prefer them to stay at home so as not to expose them to society.

This kind of behaviour that some parents still have can be very damaging for their students, because as seen in Vygotsky's theory, visual impairment interferes with a person's social life and makes them develop compensatory mechanisms so that they can fit into a society of sighted people. If the person is isolated from the world, they end up not creating these mechanisms and not developing their full potential.

With this relatively high number of DV students, it is important for the school to have adapted materials at its disposal, such as Braille or enlarged books, soroban[1] for counting, thermoform materials when necessary, among others. Miria said that these products are available in almost all schools, however, due to a lack of information, some don't request them in time and students start the school year without access to the curriculum due to a lack of adapted material.

When including students with special needs in mainstream education, it is necessary for educators to have a minimum of preparation and information on the subject, but the SEED representative said that in this administration few courses have been offered in the area of visual impairment. However, it is a positive point for schools that there are a large number of teachers in AEE, which covers all these students.

For Miria, the biggest obstacles the government faces when including students with special needs in the mainstream are the lack of accessibility and the attitudinal barrier, often due to the fact that educators don't know the needs of the students included.

BACKGROUND.

In the area of Inclusive Education, according to Carvalho (1999), the removal of barriers has been examined from the perspective of accessibility, architectural barriers, and from the psychological perspective, attitudinal barriers, which would be the lack of acceptance of these students by educators, due to a lack of preparation. Barriers to learning are obstacles that are imposed on students, generating learning difficulties. For this author, some of these obstacles are intrinsic to the student, but most are external to them. By not knowing the student's needs, or by believing that they won't be able to follow

---

[1] It's a different kind of Japanese abacus, with only five beads or pebbles in each numerical order. Its use has undergone a series of refinements that have generated extremely fast techniques for performing any calculation: addition, subtraction, multiplication, division, square root and others. (Source: http://www.sorobanbrasil.com.br/)

the process in the same way as others, the teacher can unconsciously create this attitudinal barrier. By considering that all students have different needs, trying to make learning interesting and useful for them, getting them to participate more and being more flexible, the teacher can find a way for learning to be successful not just for students with special needs, but for everyone, the way it should be.

It is believed that the use of computerised or digital teaching resources, as well as experiments involving chemical transformations, is important in the teaching and learning of chemistry. Therefore, developing or adapting these resources so that they can be used by students with visual impairments (ADV) promotes more effective and meaningful learning for them, while at the same time allowing for greater integration and inclusion of these individuals in regular basic education classes.

Computer programmes have proved invaluable for teaching chemistry, so their adaptation, combined with experimentation, could be another tool for teaching and including visually impaired students.

In view of these discussions, the problem of this research is: What is the importance and effectiveness of adapted teaching resources and experimentation in teaching chemical reactions to visually impaired students?

Main objective

To design, develop and verify whether adapted computer-based teaching resources associated with experimentation, with a multi-sensory approach, are efficient in learning concepts related to chemical reactions through ADV.

Specific objectives

- Design experiments involving non-nocuous chemical reactions that release or absorb heat and/or produce gases that are perceptible to all the senses, not just sight;
- Create an adapted teaching resource that helps ADVs understand what happens at the submicroscopic level during a reaction;
- To develop a digital educational object accessible to the blind, which will serve as an assessment tool for the activities described above.

The first chapter presents a brief history of how Brazilian legislation has evolved on the issue of inclusion, especially for people who are blind or have low vision, and how the state of Paraná plans to organise pedagogical work and the school curriculum to accommodate these students. It also discusses the theoretical framework of this research and the current state of research into teaching chemistry to ADVs.

Chapter 2 discusses the main obstacles and difficulties that students and teachers have faced in studying transformations of matter, which concepts are key to understanding the content of Chemical Reactions, and how practical activities can contribute to learning this content. It is also dedicated to referencing what has been done in the area of Chemistry Teaching, focussing on special needs, and the adaptations necessary for the teaching-learning process of this subject to be effective for ADV and visually impaired students[2].

Chapter 3 presents a review of the literature on the use of information and communication technologies in teaching, especially in science teaching, and what has been developed in the area of inclusion for the visually impaired, portraying how there are still few resources aimed at this area of knowledge.

Chapter 4 talks about the methodology adopted in the research, the subjects involved, followed by Chapter 5 which describes the materials developed and tested.

Finally, Chapter 6 shows the application of the materials and the results obtained, the conclusion of the research and the prospects for future work.

---

[2] Students who are not visually impaired, i.e. who have normal vision, are considered to be sighted

CHAPTER 1

VISUAL IMPAIRMENT

This chapter provides an overview of Brazilian legislation on special education and inclusion, from 1854 to 2007, discussing some of the obstacles that have been encountered in realising the inclusion process.

This is followed by a presentation of some data from SUED/SEED Instruction No. 020/2010, which deals with guidelines for the organisation and operation of Specialised Educational Assistance in the area of Visual Impairment in the state of Paraná. It also shows how teaching chemistry to the visually impaired should be in line with the state's Teaching Curriculum Guidelines (DCE).

Next, we discuss the theoretical framework of this research, which is based on Vygotsky's socio-constructivist theory and the zone of proximal development, as well as the multisensory didactics for teaching natural sciences, researched by Soler.

Finally, it's about the current level of research into teaching chemistry to ADVs.

1.1 History of Brazilian legislation

According to Hontangas (2010), special education is that type of teaching exclusively for people who, for psychological, physical or emotional reasons, are unable to adapt to mainstream education. This same author shows that this concept is falling apart, since all education should be special, i.e. adapted to the individual rhythm of each student.

At first, the term integration was used, which in reality, according to Hontangas (2010), ended up being confused with the simple presence of students with special needs in the mainstream school. This gave rise to the term inclusive education, which refers to education for students with or without special needs, with equal conditions so that everyone can develop their potential, respecting their individualities.

To summarise, according to the National Policy on Special Education from the Inclusion Perspective (2008), the evolution of the issue of special education has had the following history:

❖ 1854 - Creation of the Imperial Institute for Blind Children in 1854, now the Benjamin Constant Institute - IBC;

❖ 1961 - National Education Guidelines and Bases Law - LDBEN, Law No. 4.024/61;

❖ 1988 - Federal Constitution demands equal access to and permanence in school;

❖ 1990 - **Article 55** of the Child and Adolescent Statute (ECA), Law no. 8.069/90, **states that "parents or guardians have the obligation to** enrol their children or **wards in the regular**

school system";

- ❖ 1990 - World Declaration on Education for All;
- ❖ 1994 - Salamanca Declaration;
- ❖ 1994 - National Special Education Policy, stating that pupils who are able to follow and develop the curricular activities programmed in ordinary education at the same pace as other pupils should attend mainstream schools;
- ❖ 1996 - Current LDB;
- ❖ 1999 - Decree No. 3.298, which regulates Law No. 7.853/89 on the rights of people with disabilities;
- ❖ 2001 - The Guatemala Convention (1999) is promulgated in Brazil by Decree No. 3.956/2001;
- ❖ 2001 - The National Education Plan (PNE), Law No. 10.172/2001, points to the deficit in the provision of enrolments for students with disabilities in ordinary regular education classes, teacher training, physical accessibility and specialised educational care;
- ❖ 2002 - Resolution CNE/CP No. 1/2002 decrees that higher education institutions must provide, in their curricular organisation, teacher training geared towards attention to diversity and which includes knowledge about the specificities of students with special educational needs;
- ❖ 2002 - MEC Ordinance No. 2,678/02 approves guidelines and standards for the use, teaching, production and dissemination of the Braille system in all teaching modalities;
- ❖ 2003 - The MEC implemented the Inclusive Education Programme that year;
- ❖ 2004 - The Federal Public Prosecutor's Office publishes the document The Access of Students with Disabilities to Regular Schools and Classes;
- ❖ 2006 - The UN approves the Convention on the Rights of Persons with Disabilities;
- ❖ 2007 - the Education Development Plan (PDE) was launched, focusing on the training of teachers for special education, the implementation of multifunctional resource rooms, architectural accessibility of school buildings and access and permanence of people with disabilities in higher education;

The fundamental step in making inclusion a reality was the World Conference on Special Needs Education in 1994, which produced the Salamanca Declaration. This document emphasises the importance of education for diversity, since no child is the same as another and the school must exploit the potential of each one. It also demands that governments adopt the principle of inclusive education in the form of a law or policy, enrolling all students in mainstream schools and providing resources for the training of education professionals in their continuing education, as well as giving the highest political and financial priority to improving their education systems so that they are able to include all people of school age, regardless of their individual differences or difficulties.

Specialised care is not ruled out, but is provided according to the individual needs of each student, in parallel with regular school education.

But the reality is quite different, and continuing training aimed at inclusion is incipient, and in our country it goes without saying that the highest political and financial priority is not education, whether it is inclusive or not.

The Salamanca declaration (1994) advocates that educational systems adopt a pedagogy centred on the student, in other words, one that meets their needs, whether they are special or not. The challenge is indeed great, but it would be ideal, given that many students have learning difficulties. This pedagogy could reduce dropout and failure rates, because instead of highlighting the student's difficulties, it would value their potential.

Inclusion and valuing differences is a major challenge for education systems. According to the same document, the establishment of person-centred schools is a crucial step towards changing discriminatory attitudes, creating welcoming communities and developing an inclusive society.

According to the LDB, the state must guarantee free specialised educational care for students with special needs, preferably in the regular school system (BRASIL, 2010). In recent years, schools have been welcoming students with a wide range of special educational needs into their classes.

Article 59 of Chapter V of the National Education Guidelines and Bases Law (LDB) states that education systems must ensure specific curricula, methods, techniques, educational resources and organisation for students with special needs, to meet their needs, as well as teachers for specialised care and regular education teachers trained to integrate them into ordinary classes.

However, the sole paragraph of this same article specifies that students with special needs must be cared for in the regular school system, even if the aforementioned recommendations are not met. The result of this is that these students have been enrolled in many schools, but the teachers have not been trained to receive them and adapt their methodology according to these specificities. According to the Salamanca declaration, the appropriate preparation of all educators is a key factor in promoting progress towards the establishment of inclusive schools. Specialised care is also not effective in all institutions, which ends up creating an obstacle to the teaching-learning process for students with special needs.

The MEC's (2013) operational guidelines for specialised educational care in Basic Education, in the form of Special Education, define that the real intention of this care is not to replace schooling, but to run parallel to it, complementing the education of these students. As a priority, it should be carried out in the school's own multifunctional resource room, or in another regular education school, in the opposite shift to the school day, and it can also be carried out in a Specialised Educational Care centre in the public network or in non-profit community, denominational or philanthropic institutions that have agreements with the Department of Education.

Since 2010, the enrolment of students with special needs has been double-counted (in the regular school and in specialised care) in the Basic Education Maintenance and Development Fund (FUNDEB). In this way, the resources for specialised care are available if the school guides and enrolls these students in the after-school hours.

1.2 Caring for DV students in the state of Paraná

In the state of Paraná, these students are cared for in multifunctional resource rooms[3] type II and/or the Centre for Specialised Educational Care (AEE) in the Area of Visual Impairment - CAEDV. These operate in regular education establishments or in non-profit institutions, in agreement with the State Department of Education (SEED), on the opposite shift to the school day.

The minimum workload for this service is 20 hours, and in order to work in the resource room the teacher must have a degree in special education, whether or not combined with a degree in early childhood education or the initial years of primary education, a postgraduate course specifically for special education or special pedagogical complementation programmes.

According to Instruction No. 020/2010, when a student enrols in the AEE, a registration form is filled out, which includes clinical data, the need to use glasses, magnifiers, teleloupes or other items, whether they have already received specialised care and for how long, their visual acuity, visual and tactile performance and their physical mobility. To guide the organisation of the school curriculum, the state of Paraná uses the DCEs (basic education curriculum guidelines). According to this document, a subject is the fruit of their historical time, of the social relations in which they are inserted, and also a singular being, who acts in the world based on how they understand it and how they are able to participate in it. Therefore, considering the particularities of each individual is part of the teaching-learning process and inclusion is something necessary and natural.

It can also be seen in the DCEs that, for some time now, Brazilian legislation has been recommending a reorientation of curriculum policy with the aim of building a fair and egalitarian society. In other words, access to and permanence in school is the right of any person of school age, without segregation into special classes, and with equal learning opportunities for all.

The guidelines advocate a curriculum based on the scientific, artistic and philosophical dimensions of knowledge, because in this way some students who in their lives would not have access to some kind of culture and information will always have a totality of knowledge at school, and their relationship with everyday life.

---

[3] Multifunctional classrooms are spaces where a teacher with continuing training in Special Education carries out ESA. They consist of furniture, teaching materials, pedagogical accessibility resources and specific equipment and are located in schools that have enrolments of special education students. (Source: INEP)

Bertalli (2010) draws attention to the problem of welfarism and the habit of approving students with difficulties because they don't know how to work with them. Therefore, when blind students are **just "pushed" from** grade to grade, without really learning the content taught in the area of science, their education ends up being deficient, and their right to equality is deprived. It is therefore up to the school system to ensure that learning is effective for all students, in all areas of knowledge.

1.3 State of the art

According to the 2010 school census, there are 75,289 visually impaired students enrolled in the regular education system in Brazil, of whom 6,274 are blind and 69,042 have low vision. Research in this area is therefore necessary.

A person is considered blind if the corrected vision of the best of their eyes is 20/200 or less, i.e. if they can see at 20 feet (6 metres) what a person with normal vision can see at 200 feet (60 metres), or if the widest diameter of their visual field implies an arc no greater than 20 degrees, even though their visual acuity in this narrow field may be greater than 20/200. A person with low vision or low vision would be someone who has a visual acuity of 6/60 to 18/60 (metric scale) and/or a visual field between 20 and 50°. (BRASIL, 2004)

Pedagogically, the blind are defined as those who, despite having low vision, need instruction in Braille (a system of writing with raised dots) and those with low vision are defined as those who read enlarged printed matter or with the aid of powerful optical resources.

Nowadays, the school is no longer seen as just a cluster of people acting more or less in isolation, but as an interactive community, as Silva and César (2005) point out. As such, teachers are challenged to find the best way to differentiate their students and find a methodology that is effective for everyone, according to their possibilities and limitations.

Hontangas (2010) reflects that the concept of special education is no longer used, as all education must be special and adapt to the individual rhythm of each person, as each student is different, and these differences must be taken into account if the teaching-learning process is to develop fully.

Omote (2006) says that with the popularisation of discussions on inclusion, terms such as diversity and differences or individual differences have become commonplace in different situations in the daily lives of many people, not just professionals in Special Education or Education, but also professionals in other areas or ordinary citizens. But it should always be remembered that the term inclusive education defines educational movements that seek to reduce the processes of social exclusion in which many students find themselves, because they are at a socio-cultural disadvantage or because of particular characteristics such as gender, language, culture, etc.

For Ross (2006), it is important not to allow forms of assistance, protection and annulment of people with special needs, but on the contrary, to encourage and challenge them would be the right way forward, because the only way to generate discomfort and the need to overcome disability is through mediation, interaction, confrontation and action. Therefore, the teacher's role is to mediate knowledge, recognising the individuality of each student and challenging them to overcome obstacles, change their reality and improve their conditions.

However, educators are often unprepared for these changes, because the traditional teaching model massifies students without taking into account the particularities of each one. For Silva and César (2005), it can still be seen that some pupils are led into school and social exclusion, with inequality prevailing instead of inclusion and integration.

According to Hontangas (2010), the aim of education is for individuals to achieve human development and the necessary preparation to integrate personally, socially and professionally. In this way, the lack of adaptation in schools or learning difficulties cannot be a justification for segregating people with special needs in another form of education.

Bertalli's (2010) observation is pertinent when he points out that some teachers, due to a lack of preparation, end up ignoring the presence of students with disabilities and giving them token marks so that they can go on to the next grade. This leads to a certain amount of accommodation on the part of the ADVs, who have been used to this since pre-school and end up being content to carry on with poor learning.

Ross (2006) criticises the way inclusion has been based in Brazil, based only on economic equalisation, social justice, health, human rights, ecological awareness and so on. These factors are important, but they are not the basis of inclusion; the main point would be to value its challenges and objectives. Defending the acceptance of what is different would lead to the elimination of differences, but in reality it should highlight and value differences even more, the importance of listening and learning from others and the idea that everyone contributes to society in their own way. Differences are therefore not mutually exclusive, but complementary.

Hontangas (2010), for example, comments that for many students with disabilities, integration into mainstream education constitutes a subsystem of special education within the mainstream school, which gives way to more subtle forms of segregation. This is why the term inclusion, which addresses the needs of all students, is used instead of integration, which emphasises students with special educational needs.

When receiving students with special needs in mainstream education, schools have not been prepared, teachers and staff have not been trained to meet these specific needs, so it is still difficult to promote real inclusion in the classroom. It would be important to emphasise to teachers that students with special needs have learning difficulties, just as any student does, and that they all have potential that can be exploited to facilitate their learning.

> The child with special needs is not an ontologically disabled child, but a child like

all the others, with specific learning characteristics. They are not a child marked by deficits, but someone who brings together a series of attributes that can have a favourable impact on meaningful and effective learning (BEYER, 2006, p.9).

Vilela-Ribeiro and Benite (2010) researched training for inclusion in chemistry degree courses. They emphasise that this type of study not only serves to comply with public policies, but also to train reflective teachers who are capable of interpreting, understanding and questioning. Their research found that the majority of teachers relate inclusive education only to disabled students, and not to education aimed at the individual needs of all groups of people excluded from school. The undergraduates indicate that they accept inclusive education, but they don't feel prepared for it, and the probable cause is the lack of preparation on the part of the university teachers themselves.

Also according to Silva and César (2005), in order to meet the needs of all students, especially those with special educational needs, it is essential to implement methodological strategies, i.e. curricular adaptations, to better favour learning. These adaptations require appropriate scientific, didactic and psycho-pedagogical training, as well as training related to blindness and low vision. However, the latter is often lacking, which jeopardises the educator's interaction with these students. It's important not just to have institutionalised training or ongoing training through postgraduate courses, specialisation courses or training activities, but broader training through experience and collaborative work between the agents of the educational community.

Nascimento, Costa and Amin (2010) believe that in no other form of education are didactic resources as important as in the special education of visually impaired people, as they have difficulty getting to grips with the physical environment and lack suitable material. Like other students, they need motivation to learn, which can be done by taking advantage of tactile perception and facilitating the discovery of details.

The Salamanca declaration also discusses the importance of research in the field of education for inclusion, so that good examples and innovations are disseminated in schools and promote effective change in the system.

Various authors have discussed the need to develop adapted teaching materials for people with special needs, as this allows them to participate in the classroom context, to be effectively included, and to reduce their dependence on other people when they are outside the classroom.

In a study by Mól et al (2010), it was noted that there are still few dissertations and theses related to science teaching and the inclusion of visually impaired students. The authors point out that there is a need for more studies in this area, with teachers as the active subjects of these studies.

Gonçalves et al. (2013) emphasise the current state of research into teaching chemistry to the visually impaired, pointing to the low number of reports of proposals for activities in this area in both international and national science teaching literature.

The lack of a database on this subject is possibly due to the fact that it is a relatively young area that is still in the consolidation phase. Until 2010, very few articles were found in the area of teaching chemistry to ADVs, but there is currently an increase in the number of research groups.

1.4 Vygotsky's theory and visual impairment

Lev Vygotsky pioneered the notion that intellectual development occurs as a function of social interactions and living conditions. Having had experience in teacher training, he also ended up dedicating himself to the study of learning and language disorders, the various forms of congenital disabilities, and was one of the founders of the Institute for the Study of Disabilities.

For him, basic cognitive activities occur in accordance with the individual's social history, and are thus a product of the social-historical development of their community. This means that cognitive abilities and the structure of thought are not congenital, but the result of social interactions and the culture in which a person develops.

In this way, it is believed that isolating communities of pupils in special schools could be detrimental to their development, as they would be deprived of the coexistence and social interactions that exist in mainstream schools, something that from the point of view of the theory of the zone of proximal development would be unproductive.

The zone of proximal development, according to Vygotsky (1991), would be the functions and activities that the student is only able to carry out if there is help from someone, be it parents, teachers or peers. Therefore, the social nature of learning is emphasised, resulting in the fact that children develop their intellect within the intellect of those around them.

In this way, when the group is heterogeneous, learning conditions are favoured, as more advanced students can help their peers develop their potential. This implies focussing the teaching-learning process not on the cognitive functions that already exist, but on those that are developing. In other words, the student has the potential to develop knowledge, but needs help to achieve it.

Interaction through the zone of proximal development doesn't create a competence, it helps to identify its existence and develop it. Students need to interact in their environment in order for learning to trigger internal development processes. Vygotsky believes that the student's personal history and the society in which they develop are crucial to their way of thinking and learning. Nowadays, much has been studied about this issue, which has proved to be really important in the field of teaching.

The interaction of ADVs both with their peers and with sighted pupils is essential for a complete education that provides them with equal opportunities to learn, building knowledge in a rich way. In the same way for other students, socialising with disabled people of any kind (physical, visual, hearing, mental, etc.) provides differentiated learning, both in terms of school knowledge and socio-cultural issues.

According to Vygotsky (1994), thinking depends on the fullness of life, the needs and interests, the inclinations and personal impulses of the thinker. Therefore, taking into account the interests and, above all, the needs of students is important for pedagogical practice. Developing and adapting curricula and methodologies that seek to remedy each individual's difficulties is a challenge for teachers, but it must not be overlooked, as it is extremely important for improving the quality of teaching.

In Fundamentals of Defectology (1997), Vygotsky explains that in the old school, quantitative studies were carried out to determine disabilities, which were treated only as biological defects, leading to a special education that was reduced in content and slower, considering the child as different and incapable.

There was also the misconception that people with disabilities in one sense had a biological compensation in another, i.e. if you were blind, you had super hearing and touch, unlike other people. There were even religious beliefs that the person was cursed, or that they were blind because they had a special God-given power of inner sight.

In the modern conception, defended by Vygotsky, the disabled person is not considered less developed, they just develop in a different way. Therefore, like all human beings, they are individuals, different from others. Adaptations to teaching methodologies and resources, adopting a student-centred pedagogy, are necessary in any classroom, even if it doesn't have disabled students, because there is so much heterogeneity.

With more in-depth studies in the field of psycho-pedagogy, it has been realised that disabled children develop certain compensatory mechanisms, so that in order to carry out tasks there is a reorganisation and re-adaptation of the body's functions. This may not always be successful, but even if the child doesn't succeed in trying to compensate for something, they will still have some development. What we can't do is allow people to feel sorry for themselves, as this could discourage them from overcoming their limitations.

Vygotsky always emphasises that blindness is much more of a social disability than a physical one. In other words, blind people see the world differently and, in order to adapt to a society of sighted people, they develop certain compensatory mechanisms. The author goes so far as to say that if we had a group made up only of blind people, a new type of man would be created. As the education of these people is not related to blindness in itself, but to the social consequences it creates, the more they demand and the less welfare they receive, the better their development will be.

In this way, the best way to teach a visually impaired student is not to reduce the content, but to adapt it in order to make the most of their abilities. After all, blind people don't have supernatural hearing or touch, but they do develop their memory very well, adapt to their surroundings, pay close attention to what they hear, and these are skills that can be put to good use by the teacher and also in learning chemistry.

1.5 Multisensory teaching

Miquel-Albert Soler Marti researched the teaching of natural sciences to ADVs and wrote the book Didática multisensorial das ciências (1999), describing some activities and phenomena that could be observed in physics, chemistry and biology classes, making use of all our senses, without being limited to sight. According to the author himself, multisensory didactics is

> ...a pedagogical method of general interest for the teaching and learning of experimental and natural sciences, which uses all possible human senses to capture information from the environment around us and interrelates this data in order to form complete and meaningful multisensory knowledge. (SOLER, 1999, p.45)

This researcher points out that science teaching has usually been focussed solely on the visual perspective. As a result, a great deal of non-visual information is lost, ADVs don't feel motivated to study, a minimalist perception of the environment is created and a distorted interpretation of the phenomena taking place. This is why a multi-sensory approach is essential for scientific observation.

As seen above, blind people are not limited or incapable of carrying out any task, they just need it to adapt to their communication channels with the external environment, which further reinforces the idea of multisensory teaching. In addition, this type of approach can produce more meaningful and complete learning of the content, as non-visual information can also be perceived by normal-sighted students, reinforcing the concepts learnt.

Soler, as is often quoted, discusses how curiosity is often repressed in childhood, causing students to lose the habit of touching things, because when they do they are reprimanded by adults. Reinforcing curiosity in order to hear, smell and touch the things that students observe, with the necessary precautions, will make it easier for ADVs to make observations and will instil curiosity in non-blind students too, which is a positive point for a science lesson.

When carrying out activities that are strictly visual, there is the possibility of allowing the sighted pupils to give a verbal description to the ADVs. This will not only help the blind pupil to understand what is happening, but will also reinforce to the sighted pupil the details of what they are observing and increase the concentration of more dispersed pupils.

This author agrees with many others when he says that when working with this type of student, it is necessary to adapt the teaching methods used so that the input of information from the environment is equal for everyone.

In chemistry, Soler even cites examples, such as touch to explore three-dimensional atomic models, something that was used in this work; the ear to determine pH, when an acid comes into contact with some metals producing effervescence, and the sense of smell that can help identify some chemical substances.

With his research, this author drew some important conclusions that can guide the work of teachers who have DV students in their classes. For example, tactile or auditory perceptions give rise to meaningful learning for students, as long as they are accompanied by verbal explanations related to what is being observed, so mediation by the teacher or peers cannot be left aside. Furthermore, these perceptions cannot be achieved through any other sense and are of great importance to blind and partially sighted students.

Another important observation is that you can't use two perception channels at the same time if the stimuli aren't interconnected or complementary.

Auditory and tactile perceptions are extremely important for blind people, as they allow them to create mental images and make abstract comparisons between them. In learning chemistry, this is a very important skill not only for ADVs, but also for blind people.

In this work, Soler already indicated that this type of methodology would favour interaction and inclusion, since blind students help sighted students to observe non-visual characteristics and vice versa. The following is a more detailed presentation of how to work with the senses of touch, hearing, smell and taste in chemistry lessons, using Soler (1999) as a reference.

1.5.1 Touch

Through this sense you can perceive shapes, textures, sizes, weights, volumes, pressure, hardness, densities, analyse models, etc. Graphs and figures in high relief can also be very useful.

Unlike vision, which captures scientific data as a whole, touch allows you to observe in parts and then form a mental image of the whole. Remember not to provide bad sensations, as this will discourage people from wanting to observe tactilely throughout their lives.

For chemistry, the main function of touch would be to observe three-dimensional models. This didactic strategy was used to allow blind and sighted students to observe the rearrangement of atoms in molecules during a chemical reaction. In addition, mixtures involving heat exchange with the environment were made, in which the students could feel the beaker heating up or cooling down.

1.5.2 Hearing

For both sighted and visually impaired students, audio in class can serve as a motivator, a stimulus to observe nature in other ways, develop auditory attention skills, perceive their surroundings through sounds, among other things.

According to Soler, sound shows scientific concepts, which are known, recognised and retained by auditory memory. The practical examples shown by the author are more related to biology, but in chemistry you can understand the formation of gases through effervescence, as well as taking advantage of this auditory memory to help students retain important information.

1.5.3   The sense of smell

Like hearing, smell is a sense of global perception of compound stimuli, i.e. through it you can perceive a single stimulus, which is the result of a chemical mixture of others that you can't differentiate separately. Its importance is recognised because there are still no technological devices capable of capturing and reproducing smells.

When using this sense to identify chemical products, extreme care must be taken, as there are many toxic and dangerous substances, and students must be warned that they can only smell if the teacher tells them to, and not directly with their nose, but by bringing the vapours closer with their hands to smell the odours.

The sense of smell allows us to identify and differentiate one substance from another, to perceive acidity or alkalinity, to identify the products of chemical reactions, and can be useful in a variety of circumstances.

1.5.4   Taste

Soler indicates the use of taste when talking about acids and bases, to show the difference in flavours, and only this example.

This sense was not used in this work, because working with chemical substances is generally too risky. Taste could be used, for example, in research aimed at making products for human consumption or even in the wine industry, where this sense is very important in quality testing.

1.6   Teaching chemistry and visual impairment

Studying chemistry at a microscopic level requires students to articulate ideas and concepts at a high level of abstraction. Pires (2010) believes that in this case, the ADV does not present any greater difficulties than those experienced by normal-sighted students, so if they have access to the information presented at the macroscopic and representational levels, they will be able to appropriate the models and theories of chemistry just as well as other students. In this way, methodological adaptations for students with special needs can also be useful for other students.

Pereira et al. (2009) point out that there are some difficulties in teaching chemistry to the visually impaired, including the transmission of concepts strongly linked to visualisation. These authors also propose strategies to overcome these difficulties, such as research and the use of non-visual experiences that can facilitate the transmission of concepts normally anchored in visual data; the use of computer resources such as navmol[4] , the development of activities that allow blind students to

---

[4] *NavMol* is a prototype of a molecular editor for visually impaired users, who can interact with the programme via the keyboard and voice synthesiser. It allows the user to navigate the molecule atom

mentally acquire and retain the molecular structure and reaction mechanism, relating these notions to the connectivity and three-dimensionality of a given chemical structure.

As the representational level of chemistry has many specific visual resources, Braille chemical spelling was developed for use in Brazil to represent formulas, equations and chemical symbols (MEC 2002), which can facilitate the teaching-learning process for the visually impaired. Similarly, the Braille fácil programme (http://intervox.nce.ufrj.br/brfacil/), created by the Benjamin Constant Institute (IBC), which allows texts to be typed into Word by changing the font to Braille, can be a very useful tool.

One topic that has been and continues to be widely researched is experimentation in chemistry teaching. It is recognised that laboratory practices motivate and stimulate students' interest, promote the construction of various concepts and intensify the learning of scientific knowledge. In the case of ADV, Pires (2010) recognises that these activities should be adapted, valuing touch, smell, hearing and in some cases taste.

When we have blind students, it is important to carry out the experiments with a multi-sensory approach. Soler (1999) points out that a verbal description can be given of the experiment and what is observed during its execution, so that not only will the blind student have a better idea of what is happening, but also the sighted student will pay more attention to details that they might not otherwise focus on. The same author emphasises the importance of opening the doors to all the senses, not just sight, during observations, in order to make richer analyses than we usually do.

IBC provides some custom-made materials in thermoform, a kind of plastic that can be moulded from a matrix, which speeds up the process of making raised materials such as tables, graphs, figures and diagrams that allow the blind to follow tactilely. For chemistry lessons, there is the atom model, the periodic table, the electronic distribution booklet and the periodic properties table.

In the scientific literature, some materials have already been created for use by ADVs, such as a card game that shows some chemical properties like nox, a domino of organic substances, some atomic models, various periodic tables, a molecular model for working with hydrocarbons, a biscuit molecular model for working with various structures and thermochemistry graphs in relief and different textures.

---

by atom, providing information about neighbouring atoms and bonds, and can add or delete atoms and bonds if desired. (http://www.molinsight.net/)

CHAPTER 2

CHEMICAL REACTIONS

This chapter aims to refer to and discuss the main difficulties secondary school students have in understanding concepts related to chemical reactions, such as alternative conceptions and epistemological obstacles. It then goes on to explain the suggestions given by some authors for solving or circumventing the main difficulties.

Finally, we discuss the state of the art in the study of chemical reactions in classrooms with blind students, both nationally and internationally.

2.1 Concepts and methodologies of chemical reactions

Transformations of matter are one of the focuses of the study of chemistry, since the other fields of this science are directly related to chemical reactions (speed, equilibrium, stoichiometry, electrochemistry, etc.). A correct understanding of the concepts of the transformation of matter involves the three levels described in Johnstone's famous triangle represented in figure 1 (1993), an idea supported by Brazilian researchers such as Mortimer (2000).

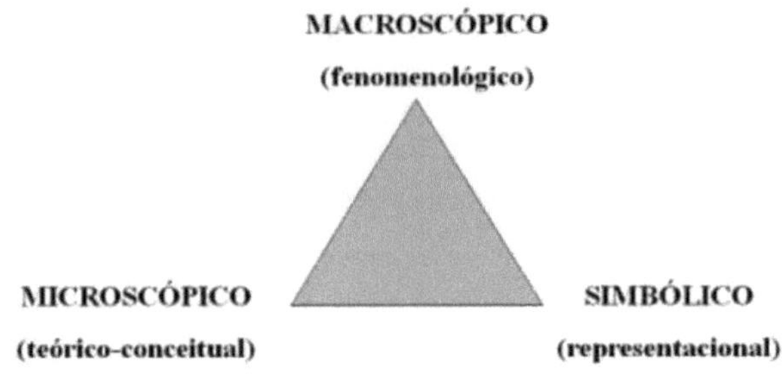

Figure 1 - Triangle of meanings in chemistry Source: Canzian and Maximiano (2010)

At the phenomenological level there are empirical observations of transformations in everyday life or in the laboratory, in which the macroscopic aspects of such transformations are easily observed. At the theoretical level there are descriptions of phenomena, explanations at the atomic-molecular level, which require greater abstraction on the part of the students, often constituting a barrier to understanding the content. Finally, the representational level involves transcribing the observed phenomenon into equations.

The key concepts are the identification of chemical reactions, their understanding as the rearrangement of the atoms of each substance involved, the conservation of mass and the representation of these phenomena in the form of equations.

Several authors agree that the study of chemical reactions is one of the pillars for understanding the organisation and construction of chemical knowledge. However, the approaches normally adopted for teaching this content have led to a number of learning problems because they don't adequately deal with the numerous abstract notions that make up such content. Many textbooks, for example, still use criteria such as reversible and irreversible or macroscopic changes in the system to classify and differentiate between chemical and physical transformations.

Mortimer and Miranda (1995) consider chemical reactions to be a central concept for learning chemistry, as they cover a range of content, since their understanding depends on recognising that matter is made up of atoms and that these atoms are conserved in these transformations. They emphasise that secondary and primary school students find it difficult to study chemical reactions due to the great extent and generality of this concept. There is often confusion between a change of physical state and a chemical transformation and obstacles to recognising the reaction as an interaction between substances.

In addition, teachers traditionally emphasise representations to the detriment of phenomena, which can lead to students maintaining the aforementioned misconceptions, failing to relate the transformations that occur at the phenomenological level to the explanations at the atomic-molecular level.

The article by Filho and Celestino (2010) points to confusion between the terms mixture and reaction, dilute and dissolve, difficulty in differentiating between chemical and physical transformations, understanding of the concept of solution as limited to solid-liquid and classification of mixtures based only on phases. Their research also demonstrated the lack of a microscopic view of solutions.

There are studies such as those by Justi and Ruas (1997) which even show **that for students in the** "atomic world" particles change shape, size and colour, just as substances **do.**

Chagas (2007) proposes solving problem situations as a methodology for working with chemical reactions. Observing these activities, he found that the observations made by the students did not go beyond immediate intuition, with a strictly visual reading of the phenomenon prevailing, constituting the obstacle of the first experience of Bachelard's theory. He also noted the existence of the substantialist obstacle in terms of attributing qualities to substances, considering that one acts on the other and not that there is a relationship between them.

To overcome these obstacles, Mortimer and Miranda (1995) propose using discussions of the explanations that students give for some simple chemical transformations, as well as the teacher's reinterpretation in atomic-molecular terms. In this way, the teacher will promote the establishment of relationships between the observations and interpretations of the phenomenon and the explanation at the submicroscopic level.

Before representing chemical reactions using equations, it is also important to discuss the exchange of energy, the different rates at which chemical reactions occur, among other things. This introductory discussion will show that equations are a simple way of representing a much more complex phenomenon, preventing the representation from being confused with the phenomenon.

Understanding the three levels of chemistry is therefore essential for learning to be complete and effective. Knowing and interpreting the representational level makes it easier and quicker to read and write about phenomena. In the case of ADVs, Braille chemical writing (GQB) allows this level to be worked on. However, the article by Resende Filho et al. (2013) showed that the students' knowledge of chemical symbology through GQB is low, since they did not know how to represent symbols such as chemical elements, molecular formulae and chemical equations. Therefore, she concluded that the fact that they don't know GQB is one of the main factors behind the difficulties ADVs encounter in learning chemistry.

2.2 Experimentation for blind and visually impaired students

Filho and Celestino (2010) concluded that surveying previous conceptions and carrying out experiments discussed in small groups helps students' conceptual development, enabling a better understanding of the phenomena under study.

Experimentation, problem-solving and discussions emphasising the relationship between substances appear to be effective methodologies for teaching and learning chemical reactions. The transposition of the phenomenon to an atomic-molecular level must also be done carefully, so that students understand the real meaning of the representations in chemical formulae and equations.

The role of experimentation in teaching and learning the natural sciences is undeniable. This is one of the areas most investigated by renowned researchers, who recognise that it is a motivational tool for students, a way of building knowledge from concrete situations, and that it should always be linked to the learning of content, and not just as a "show" to attract students' attention.

According to Giordan (1999), the development of scientific knowledge is intrinsic to an experimental approach, since its organisation takes place preferably in the middle of the investigation. In other words, the development of hypotheses and explanations for phenomena observed during experimental investigations brings students closer to the construction of scientific knowledge.

Barberá and Valdés (1996) emphasise that experimentation needs to be planned in relation to the objectives you want to achieve, and often teachers agree that it is fundamental in science teaching, but they don't know exactly why they should use it, limiting themselves to demonstrations of consolidated theories, or just as motivation. Students know how to observe, classify and hypothesise on their own; the teacher's role is to teach them how to do this in a scientific way and in scientific language.

The same authors go on to say that these activities make it possible to contrast scientific abstraction with reality, which is much richer and more complex than theory, thus highlighting some of the epistemological obstacles that may exist, in order to replace them with scientific concepts. It also familiarises students with certain technologies and improves practical thinking. After all, the best way to learn how to do science is by doing it.

Guimarães (2009) also defends the idea that carrying out experiments like cake recipes is not the best way to stimulate learning. Practical activities can be used to test and prove hypotheses, to demonstrate a principle, or as an investigation, the latter being the one that best helps students learn, as it allows them to build up knowledge little by little, through observation, creating cognitive conflicts that are overcome, and generating meaningful learning.

Garcia Ruiz and Calixto (1999) also point out that experimental activities enable students to develop scientific thinking, acquire theoretical knowledge, stimulate them to think and verify their own observations and explanations about the phenomena that surround them, building their learning and generating more critical thinking. In addition, it allows the teacher to move away from the role of simple transmitter and become a mediator of knowledge, reflecting on the students' difficulties and how they best appropriate the information.

According to Nunes et al. (2010), little work on the inclusion of ADV has been carried out in Brazil, such as that of the UnB master's group, which has been working to improve this situation. They carried out research with undergraduate students, proposing chemical reactions and experiments that valued all the sense organs. The experiments included the production of casein glue, the study of reaction speed based on the effervescence of a tablet under different conditions, acid and base differentiation, a demonstration of how a battery works and a study of the thermal sensation of different materials. Their results showed that teaching chemistry to the blind is not an impossible task, and that by encouraging multisensory methodology all students are favoured. Chemistry is very much related to visual observations, so the important thing is to instil in undergraduates that it is possible to use the other senses to learn this science.

Gonçalves et al. (2013) carried out a chromatography experiment in a regular education class with 28 sighted students and 1 blind student, in which the students obtained a chromatogram with a marker pen and alcohol, and then adapted it for their blind classmate, using EVA and textured materials to highlight the difference in colours. The results were interesting, and the promotion of inclusion was observed, because in order to assemble the chromatograms, the sighted students needed to interact with the ADVs, to know what their perspective would be when reading the image, resulting in improved learning on the part of all the students.

In the international literature, Cary Supalo is one of the researchers with the most publications and research related to teaching chemistry to the blind. His work mainly involves laboratory techniques and adaptations.

In countries such as the United States, when carrying out chemical experiments, students have an assistant whose job it is to carry out the actions as the ADV directs, even if they are wrong. In order to help blind people observe chemical reactions, Supalo et al. (2006) developed SALS (Submersible Audible Light Sensor), which is a tool that registers colour changes or precipitate formation by emitting an audible signal. SALS also has a Mobile Color Recognizer, which allows the user to take a photo of the reaction and be informed sonically of what colour the system is at that moment. This type of initiative is very interesting because it allows ADVs to work and obtain information about the reaction in real time, interacting better with their sighted peers.

CHAPTER 3

INFORMATION AND COMMUNICATION TECHNOLOGIES

This chapter aims to discuss the importance of ICT in the classroom today, as well as the main studies carried out on the use of ICT in the teaching of science, and more specifically chemistry.

At the end of the chapter there is a review of some of the work carried out in the area of inclusion for the visually impaired, showing how there are still few resources focussed on this area of knowledge.

3.1 Use of Technology in Teaching

The role of technology in today's society is undeniable, and consequently teaching processes must keep pace with these developments in media and information technology. Public and private initiatives have invested resources in equipping schools, above all because of the current labour market's demand for computer skills. However, educational activities must go far beyond training for the job.

This process is not at all simple; on the contrary, it requires common sense, adequate continuing training and investment from the competent bodies in both equipment and human resources.

For Cysneiros (1998), the failure to use technological artefacts in teaching has followed a cycle: research pointing out the advantages, discourse on the obsolescence of the school, implementation of public policies to introduce the new technology, limited adoption by teachers and the lack of significant academic gains. Then comes the research that addresses the reasons for these failures, the defence of even better technologies, their implementation in schools, starting the cycle all over again.

The same author also argues that we are living through conservative innovation, because the special and seductive effects of media such as sound, television, computers and images are taken advantage of, but the lecture format is maintained, in other words, nothing other than the appearance of the lesson has been innovated. Therefore, mediation by the teacher is crucial if ICT is to fulfil its role and really bring significant gains to the quality of teaching.

Giordan (2005) believes that research involving the use of Information and Communication Technologies (ICTs) in education is extremely important, as it is a mediational medium that conditions teaching actions and the development of higher mental functions. Cultural tools have a great impact on human actions, so when using them in the classroom you need to know what objective you want to achieve.

In his book Cyberculture (1999), author Pierre Lévy highlights the drastic change in training for the labour market, since what we learn at university today will have become obsolete in a few decades or less. Continuing education has therefore become more essential than ever.

With the growth of information technology and the so-called cyberspace[5] , according to Lévi (1999), new forms of access to information and new styles of thinking and knowledge have been created. However, care must be taken not to fall into the illusion that access means acquiring knowledge.

Of course, it's much easier to find information of interest and participate in forums and communities that meet the needs of each individual, but the universe of what can be found on the web is so vast that filtering this information becomes an increasingly arduous task.

And it's the teachers who must help find the filters in the first place. Of course, they won't be replaced by machines, but their practices must be modified as a result of their use.

For Lévy (1999), a change is needed in the organisation of content, in levels and prerequisites, to open up less linear, more open spaces that are organised according to objectives or contexts.

This transition has been happening little by little, but too slowly. Some textbooks have already modified the structure of the content, and curriculum plans in Brazil have been trying to break down this structure into linear topics, but in practice it is still difficult to break out of this sequence that is so ingrained in the traditional school structure.

Lévy (1999) goes on to say that the role of the teacher transcends that of a transmitter of knowledge, since other means are more effective for this, and becomes that of an encourager of learning and the search for knowledge. It is the teacher who must mediate the students' relationship with so much knowledge that can be found on the Internet.

It's impossible to deny that television, and nowadays the computer to a much greater extent, influence learning, as students spend most of their time in front of a computer screen. Of course, the benefits cannot be overlooked, as they can have contact with different cultures, broaden their horizons and change their way of thinking. But for this to happen, the school and teachers are essential in pointing the way, so that technology doesn't become an instrument of alienation.

Dias (2010) emphasises the need for accessibility in education and the importance of the teacher as an active subject in these adaptations and development of accessible educational objects. More importantly, the challenge encountered when adapting digital educational objects so that they are not restricted to one group of students, in other words, that they help ADVs and visually impaired people to learn certain content, so that inclusion is real.

Eye-catching sound resources and descriptions are therefore the simplest technologies to use

---

[5] Cyberspace is considered a virtualisation of reality, a migration from the real world to a world of virtual interactions (Source: Levy, 1999).

for students with or without visual impairments. The concomitant use of resources with screen readers also helps the teacher when using objects that are not adapted for the blind.

Nowadays simulations are the most widely used teaching tools, but for VLE the iconographic characteristics, illustrations and representations in the windowed environment can be accessibility barriers. Dias (2010) found in a survey of 50 learning objects belonging to RIVED[6] that none of them had accessibility, making it impossible for people with visual impairments to use them.

In this research, the tool developed and used was what Giordan (2005) defines as a tutorial system. The student interacts with texts and some definitions, is questioned about the topic previously studied and receives feedback on their response. The aim is for both sighted and visually impaired students to be able to use the resource without major difficulties.

Carvalho et al. (2003) developed a programme, BR Braille, capable of transcribing what the student writes in Braille into the optical system, by scanning and recognising codes, thus making it easier for sighted teachers to read the work done by blind people. It was a major step forward in the development of technologies for ADV, as the proposal is also affordable.

Lopes et al (2011), arguing that access to ICT without adequate teaching and learning strategies is not enough for quality education, developed a maths learning object accessible to the blind. Initially, the project was going to be in flash[7], but screen reading is compromised because the most commonly used readers don't read images, graphs and tables, so the object was made in html[8]. Through the application, the students felt that the learning object was clearer and more objective than the explanations in the books, and more attractive and interesting in the way it approached the subjects.

Miner et al. (2001) state that computers offer a variety of resources for students with special needs, helping to reduce difficulties in accessing information and communicating with their peers and teachers. They also point out that ADVs learn chemistry better in the classroom and in the laboratory when they have access to the appropriate combination of hardware, software and other assistive technologies (products, resources, methodologies, strategies, practices and services that promote the functionality, autonomy and quality of life of people with disabilities) on the computer.

---

6 RIVED is a programme run by the Secretariat for Distance Education (SEED), which aims to produce digital pedagogical content in the form of learning objects, with the goal of improving learning in basic education subjects and students' civic education. As well as promoting production and publishing digital content on the web for free access, RIVED provides training on the methodology for producing and using learning objects in higher education institutions and the public education network. (Source: http://rived.mec.gov.br/)

7 Flash is a multimedia platform for developing applications containing animations, audio and video, widely used in the construction of commercials and interactive web pages.

8 HTML or HyperText Markup Language is a markup language that basically defines the pages we use on the web.

CHAPTER 4

RESEARCH METHODOLOGY

This chapter will describe the methodological framework adopted in this research, as well as describing the instruments used and the research subjects.

The information was gathered using different oral and written instruments. In order to observe the interactions and the development of the research, the data was recorded through video recording associated with questionnaires and written activities to evaluate the results.

4.1 Data collection through video recording

Martins (2006) points to the increasing discussion of the role of language and discursive interactions in the classroom, as well as highlighting that many authors argue that social interactions are considered essential for learning. In order to conduct these studies, the use of videotaping as the main instrument for collecting empirical material is an interesting data collection tool, since it makes it easier to identify the interlocutors, especially in large groups such as the classroom, as well as making it possible to document elements of non-verbal communication alongside verbal language.

According to Duarte (2002), video recordings are interesting in qualitative research because they allow us to observe the posture of the agents during the research, gestures and body signs that can later provide significant elements for interpreting and understanding the universe under investigation.

Carvalho (1996) observes that the use of video allows the researcher to see and review the learning episodes of interest, and can extract information that was not observed when the project was applied in the classroom. In order to extract the maximum amount of data from this type of collection, the author indicates the need to initially separate the episodes in a crude way, then try to classify them, discuss the classification made with peers, analyse the episodes and finally triangulate the data with other data obtained using other instruments.

To obtain the video recordings, the students and teacher signed a consent form. A laptop webcam was used, which captured the whole room but didn't identify the sounds well. In order to better capture the interactions of the sighted and ADV groups, an ordinary 5 megapixel camera was used, which was passed from bench to bench during the experiments to record the students' actions. After the questionnaires had been administered, a recorded interview was also conducted with the ADVs, so that they could explain what they thought of the work, what the positive and negative points were, and what suggestions they had for improving the products.

4.2 Qualitative research

According to Silva and Menezes (2005), qualitative research considers that there is a dynamic relationship between the real world and the subject, in which this link between the objectivity of the world and the subjectivity of the subject is inseparable. In this way, translating the results of social research into numbers would be impossible.

In this type of research, the interpretation of phenomena and the attribution of meanings are the real basis, and the use of statistical tools is not necessary. As Silva and Menezes (2005) describe, the natural environment is the source of data collection and the researcher is the key instrument.

According to Bogdan (1994), in qualitative research the source of data is the natural environment, and the researcher is the main instrument. Qualitative researchers are more interested in the process than simply the results or products.

For Rey (2002), qualitative research emphasises the condition of the researcher as a subject and the importance of their ideas in producing knowledge. This knowledge is not the immediate proof of the sum of the episodes, but has an interpretative character, the aim of which is to give meaning to the expressions of the participants studied.

Patton (1986) indicates three characteristics that he considers essential to qualitative studies. A holistic view, i.e. it is only possible to understand the meaning of a behaviour or event by understanding the interrelationships that emerge from a given context. Inductive approach, in which the researcher starts from more free observations, allowing the dimensions and categories of interest to emerge during the data collection and analysis process. Naturalistic enquiry, in which the researcher's intervention in the context is kept to a minimum.

Considering the problem and the objectives of this research, it can be seen that the qualitative treatment of the data is the most appropriate, since the number of ADVs in a regular class is not large enough to generate statistical data, and analysing the evolution of the students and the real interaction between the blind and the sighted during the lesson is the best way to evaluate the results.

From the point of view of technical procedures, this is Action Research. According to Engel (2000), unlike traditional research, which is objective, independent and non-cooperative, action research seeks to develop knowledge as part of practice. In other words, this type of research is carried out by someone who is part of the practice and who wants to improve understanding of it, given the need to bring theory and research in the field of teaching closer to classroom practice.

Koerich et al. (2009) characterise action research as social research that is based on a collective problem in which the researchers are involved, so that they can seek solutions in a cooperative and participatory way.

4.3 Introducing the Subject and their Context

This research was carried out in a 2nd year secondary school class at a state school in Curitiba. Of a total of 27 students, 3 are blind and 1 has low vision. From now on, the three blind pupils will be referred to as pupil A, pupil B and pupil C, and the pupil with low vision as pupil D. The disabled pupils receive specialised care on the opposite shift to their schooling at the Instituto Paranaense de Cegos and in the school's multipurpose room. In this room, there is a professional who works for a period of time in the school, looking after students with special needs.

Student A lost her sight after she was born between 6 and 8 months old. She uses a laptop in class and is proficient in braille. Students B and C have been blind since birth and also use a laptop with the DOSVOX programme[9] and have mastered braille. Student D has low vision, with between 15 and 20 per cent visual acuity. He doesn't have a laptop, but has mastered Braille and can work with enlarged print, as well as writing in ink, with a little difficulty.

The pre-questionnaire applied to the students in this research was intended to check their previous conceptions of the content of chemical reactions and the learning difficulties in chemistry associated with visual impairment. It was validated by professors Dr Claudia Regina Xavier and Dr Maria da Conceição de Almeida Barbosa-Lima, as well as the supervisor, Profa Dr Fabiana Hussein. The purpose of the post-questionnaires was to analyse whether the work carried out led to better learning of the content covered and whether the software created was effective.

The post-questionnaire consisted of open-ended questions or Likert-type scales, so that when respondents gave a high value they agreed with the statement, and lower values indicated disagreement. This type of question was used only to assess the students' satisfaction with using each material proposed in the survey, and their opinions on whether or not the material helped them understand chemical reactions. The open questions, on the other hand, were designed to see whether or not the conceptions identified in the pre-questionnaire had actually changed.

Two questionnaires were used to evaluate the software. Nokelainen's (2006) evaluates criteria such as student control, student activity, goal orientation, applicability, motivation, evaluation of prior knowledge and software flexibility. The other instrument, which is attached, was an adaptation of the questionnaire proposed by Relvas (2005), to assess understanding of the interfaces, audio, lyrics and the suitability of the software for the content being worked on.

---

[9]DOSVOX is a system for PC microcomputers, developed at UFRJ, which communicates with the user via voice synthesis, thus enabling the use of computers by the visually impaired, who thus acquire a high degree of independence in their studies and work.

CHAPTER 5

DESCRIPTION OF THE MATERIALS DEVELOPED

This chapter describes the materials that were developed to work on the content of chemical reactions with blind and visually impaired students. The materials described are the manual with the didactic sequence for inclusive teaching, the magnetic molecular model, the way in which Braille chemical spelling was worked on, and the description of the Inclusive Chemistry computer programme.

5.1 Teaching materials for inclusive education

The teaching material for inclusive education consists of a sequence of activities and assessments on chemical reactions, with accessibility for blind students. Four actions are presented to help chemistry teachers who have blind or low-vision students in the classroom.

The first action aims to diagnose the learning difficulties related to the content and the students' alternative conceptions through the application of a questionnaire.

Action 2 presents the sequence of experiments with a multisensory approach to be carried out, as well as the questions that can be worked on from these practices. In these experiments, the students identify whether there is a chemical reaction or not, describing the initial and final characteristics of the systems, in order to conclude what evidence accompanies the phenomena.

The third action is the transposition of phenomena into theory, working with texts and discussions, and with the adapted molecular model, which will be presented in item 7.2. Together with this action we have the transcription of the reactions studied at the phenomenological and theoretical level to the representational level, through chemical equations, using the MEC's Braille chemical spelling and the adapted molecular model to help balance the equations.

Action 4 describes the development and justification of the content used in the Inclusive Chemistry Q.I. software, indicating how teachers can use it in the classroom.

5.2 Multi-sensory experiments

The experiments carried out in the first lesson were as follows: dissolving an effervescent tablet in water, so that the ADVs could hear the sound and feel the solution bubbling during effervescence; mixing sodium bicarbonate and vinegar to notice the change in the odour of the substances; mixing hydrochloric acid and sodium hydroxide, to show that some reactions don't show observable macroscopic changes. After each reaction, the students described the macroscopic characteristics of the initial system before mixing and the final system after mixing the reactants, indicating what evidence there was that a transformation had taken place.

In the second phase of experimental activities, the following experiments were chosen: dissolving urea in water, which, because it is an endothermic dissolution, causes the beaker to cool slightly; a mint in soda to observe that physical transformations are evident even when there is no reaction. The questions raised were what the students noticed when they dissolved the two substances, whether they believed that new substances were formed, and whether it would be possible to obtain the initial solids again by letting the water in the system evaporate. The questions for the soda experiment were what changes were observed in the system, whether the gas released was produced by a reaction or already existed in the initial system, and whether any new substances were produced in this system.

In the last experimental activity, three practices were carried out: the reaction between sodium hydroxide and copper sulphate II, in which a precipitate is formed and the mass does not change before and after the reaction; the reaction between sodium bicarbonate and vinegar, in which the mass decreases due to the loss of carbon dioxide from the system; and the burning of steel wool to observe the increase in weight due to the reaction with oxygen in the air. A digital scale was used for the sighted students, while a two-pan scale was used for the ADV students, so that they could see whether or not there was a difference in level after the reaction. The questions for analysis and discussion were whether the mass changed before and after the reaction, explaining why.

5.2 Magnetic molecular model

A magnetic molecular model has been created, with adaptations for use by blind people. It consists of Styrofoam spheres that stick to a magnet on a magnetic board of the type used to put up photos. Each ball has the symbol of the element it is supposed to represent in Braille, so that blind and sighted students can assemble the molecules of the reactions indicated by the teacher, making it easier to balance and visualise the submicroscopic level by means of a representation.

Materials: Styrofoam balls, coloured paint, a paintbrush, cardboard (or paper with a specific weight for writing in Braille), a needle (or punch and straightedge), small magnets found in craft shops and a magnetic board of the type used to stick notices or photos, easily found in stationery and gift shops.

Assembly: paint the Styrofoam balls different colours to represent atoms of different elements. Stick the magnet on the ball once the paint has dried. Using cardboard or, if the school has it, braille writing paper, write the symbols of the elements in braille, cut them out and stick them to the balls. When working with this model, the teacher must be careful to point out that the colour is only a representation, and does not mean that the atoms have such colours.

To write the symbol of the elements, according to the GQB, the capital letter symbol is used at points 4,6 of the Braille cell followed by the letters that normally represent the element, as shown in figure 2.

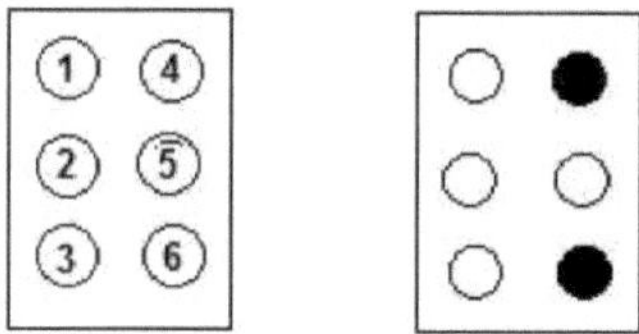

Figure 2 - Empty braille cell with the corresponding numbers for each dot and the sign indicating a capital letter

Figure 3 below shows some of the elements used and Figure 4 shows a molecule built using the molecular model produced.

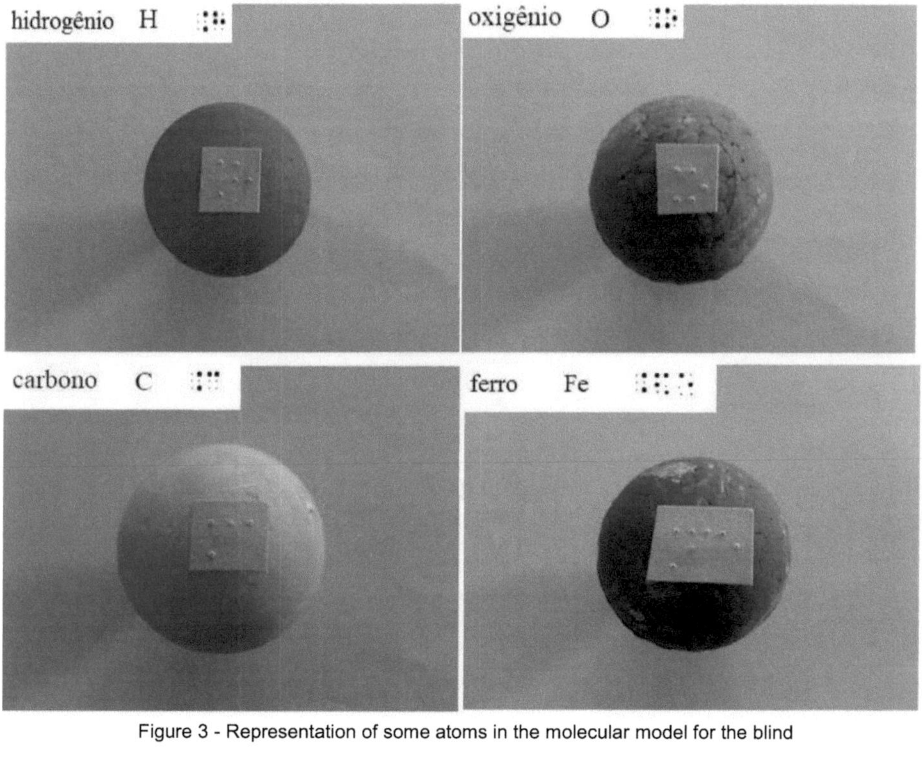

Figure 3 - Representation of some atoms in the molecular model for the blind

35

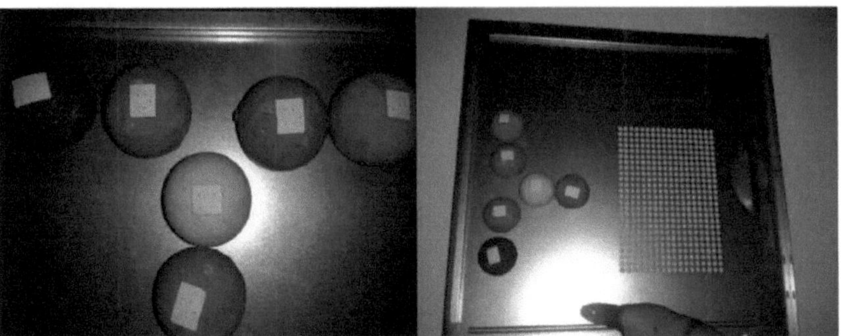
Figure 4 - model of the sodium bicarbonate molecule made using the magnetic molecular model

5.4 Braille equations

To make braille writing easier for teachers, there is a computer programme available on the Instituto Benjamim Constant website called Braille Fácil (http://www.ibc.gov.br/Nucleus/?catid= 79&blogid=1&itemid=387). When you install it, the BrailleKiama font will be available in word and the teacher can type and print the dots they want. The dots can be covered with glue to make them stand out. Figure 5 below shows an example of a reaction written in this way:

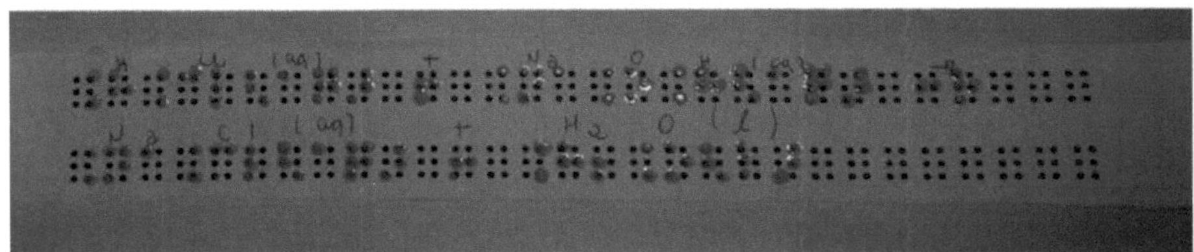
Figure 5 - Reaction made using the Easy Braille programme with 3D glue

According to the MEC's braille chemical spelling, the chemical elements in braille are transcribed according to the common system, using the normal symbols for each element, preceded by dots 4,6 to indicate capital letters.

The lower indices on the right, representing the number of atoms in chemical formulae, are transcribed at the bottom of the braille cell, with no position indicator and no digit sign.

For stoichiometric coefficients, the digit sign (3,4,5,6) is used and no empty cell is left between the coefficient and the element that follows it.

The physical states are represented by corresponding abbreviations in brackets, placed right after the substance's formula. Figure 6 shows how part of a reaction is done in GQB. Figure 7 exemplifies a complete reaction in braille.

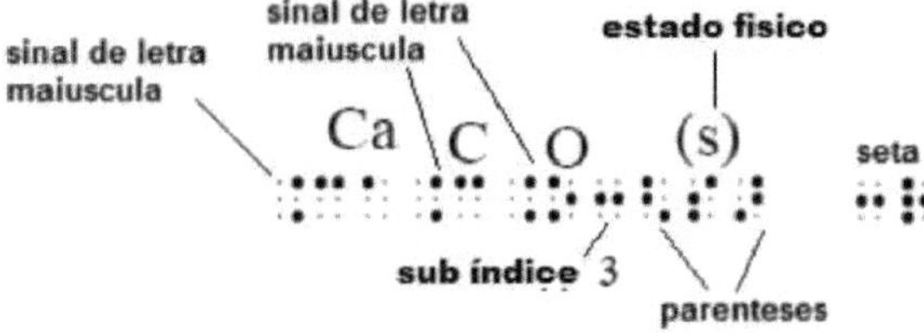

Figure 6 - Representation of part of a reaction in braille chemical writing

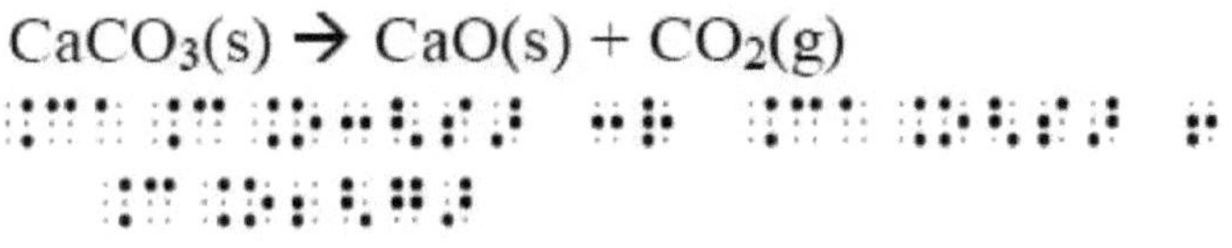

$$CaCO_3(s) \rightarrow CaO(s) + CO_2(g)$$

Figure 7 - Complete chemical reaction written in braille

5.5 Computer programme

In this master's programme, a digital educational object has been developed which is designed to assess learning after the teaching material has been applied. It is a kind of question and answer game, in the form of a tutorial system, in which the student receives feedback immediately after giving their answer to the question. This software was developed using the English programme Atoms, Symbols and Equations as a model.

The aim of this material is to help teachers with ADV to develop an activity that can be done at the same time by sighted pupils too, serving as a learning assessment tool and motivating pupils, especially as there are few digital objects that can be used by blind people.

The programming was carried out by a Computer Engineering student, André Matheus Fedalto, who is in his 7th term and is a scientific initiation scholarship holder of the professor supervising this research, Professor Dr Fabiana R.G. e Silva Hussein.

The *software* was made using the latest web development technologies: HTML5, CSS3 and *JavaScript*. This means that the *software* can be accessed both locally and remotely, with a view to providing access via the internet in the future. Another advantage of using these technologies is that they are becoming a standard across all devices, being supported by computers, *tablets* and mobile phones. This means that there is no *hardware* or operating system impediment to the *software*'s operation, as long as it is minimally recent and has a web browser installed (Internet Explorer, Google Chrome, Mozilla Firefox, etc). With HTML5, the software doesn't need *plugins* or third-party applications to work, since this technology already supports audio natively, and all navigation and verification is done via *javascript,* i.e. locally on the device it is running on, so there is no need for a server to process the data. The *software* is also already prepared and supports styling via CSS3, so any experienced *designer* can easily change the *layout of* the *software* to make it more attractive or just adapt it to their use.

All of the object's screens come with audio, so that ADVs can use it in the same way as sighted people and at the same time. First we have a welcome screen, figure 8, which shows the commands needed to navigate the programme.

Figure 8 - Inclusive Chemistry programme presentation screen

This is followed by a brief introductory text on chemical transformations, just to remind you of what you have already seen in previous lessons, provided that the teacher has worked with the sequence of activities suggested here, otherwise this software should be used after working in the classroom with the chemical reactions content. The questions and activities on chemical reactions are then worked on, taking up the concepts of definition, representation using formulae and balancing equations. In this way, the digital object covers the previous stages of the sequence of activities, including questions on each of the contents worked on in the classroom.

Figure 9 shows one of the programme's screens after the student has given the correct answer.

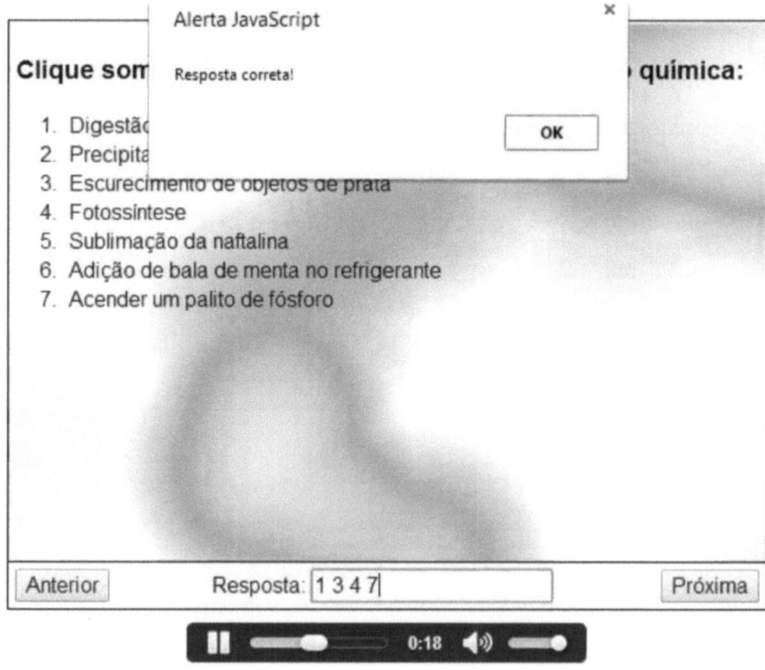

Figure 9 - Programme screen after an answer has been given

The programme interactively assesses students' level of knowledge in recognising chemical reactions, identifying substances present in some phenomena by the name by which they are usually known, and also by their formulae. It also presents questions on the classification of reactions. Further details are given in the Teaching material for inclusive education.

CHAPTER 6

RESULTS AND DISCUSSIONS

The results obtained from the application of the methods and resources described above will be presented here, along with their discussion.

6.1 Application of the didactic unit

The results obtained during and after the application of the didactic material for inclusive education are presented below, detailing each activity carried out and the effectiveness of the methods used. The application took a total of eight 50-minute lessons, and each day of the week the class had two twinned lessons with the application of this project.

6.1.1 Questionnaire

The first stage of the work is to diagnose the knowledge that all the students in the class already have on the subject. For this work, a pre-questionnaire was applied, not only to check what the students knew about the subject, but also to identify their difficulties.

The time allotted for this activity would have been one 50-minute lesson, but the DV students were a little slower than the sighted ones to read and answer the questions on the computer, so the time was one and a half lessons, i.e. 80 minutes.

The application was attended by 23 sighted students, three blind students and one student with low vision. **The first question was** "According to what you already know about the subject, what is the difference between physical and chemical transformations?" It turned out that most of the blind students (11) didn't know the difference between chemical and physical transformations. Those who answered the question didn't have a clear definition either, saying that physical transformations are those that occur naturally, while chemical transformations are those that are artificial. Only one student mentioned a phase change as a physical transformation.

Among the DV students, student A was the only one who gave a more **accurate** definition: **"Chemical transformations occur on the basis of the reactions that** take place between the chemical elements that generate new substances, while physical transformations are not based on these **reactions."**

Students B, C and D were unable to identify the differences, considering only that chemistry has reactions and physics does not. Apparently they differentiated between physics and chemistry rather than transformations. This indicates that the ADVs have a slightly better understanding of these transformations, since most of the sighted students didn't even try to write down the difference, while all the ADVs answered the question.

Regarding the identification of a chemical reaction, 11 students related macroscopic changes to the system and 2 even mentioned a change of physical state as a reaction, confirming Mortimer and Miranda's (1995) idea that students have difficulties identifying chemical reactions, as they stick to the visual aspect and even say that phase changes are chemical transformations.

Only one student said that observation is what makes it possible to identify a reaction, **while another wrote that it is the "exchange of molecules".**

Student A wrote that the reaction occurs **"when** the elements existing **before the reaction change". Student C didn't answer, and students B and D** said that when we mix products, a reaction occurs.

A list of phenomena was given to the students so that they could give their opinion on whether they represented chemical reactions or not. The phenomena in the questionnaire and the results are shown in figure 10 below.

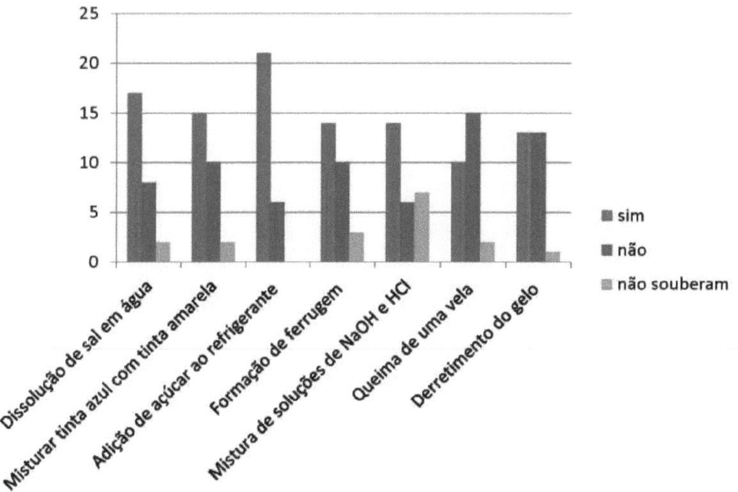

Figure 10 - Phenomena identified as chemical or non-chemical reactions

Once again, it can be seen that there is still no clear distinction between what a chemical reaction is, as dissolving salt and melting ice were mentioned many times as reactions. In addition, once again we see the obstacle of the first experience, cited by Chagas (2007) as the limitation of observation based only on visual evidence, without considering what actually happens to the atoms and molecules in the reaction. This fact is also clear when the students mention adding sugar to soft drinks and mixing paints as a chemical reaction. Even more indicative of the lack of knowledge of chemical reaction content is when many students don't consider burning a candle to be a chemical reaction.

In order to analyse the students' definition of chemical reactions, they were asked to name three reactions that take place in their daily lives. Among the DV students, student C couldn't answer, students A and B cited examples that actually corresponded to changes of physical state. Student D wrote down some formulae, such as that for water. Only student A mentioned "the appearance of fire". It was no different with the normal-sighted students: six cited situations that were given in the previous question, six didn't know how to answer, five cited physical transformations, and many cited only products that they believed contained chemicals, such as cosmetics, soft drinks, cleaning products, but not reactions as such. Only five students mentioned the fermentation of bread or cake, four mentioned lighting a fire and one mentioned adding effervescent to water.

Mortimer and Miranda (1995) also found that secondary and primary school students find it difficult to study chemical reactions due to the great extent and generality of this concept. There is often confusion between a change of physical state and a chemical transformation and obstacles to recognising the reaction as an interaction between substances. This fact was proven with this initial questioning, indicating that not only the ADVs, but all students, still find it difficult to understand this content.

The questionnaire made it clear that students have difficulty interpreting what happens at the submicroscopic level in a chemical reaction, as 12 of them (44 per cent) were unable to answer what happens to the atoms and molecules in the reaction of an effervescent compound with water. The most common answers were that it changes from a solid to a gaseous state or that it dilutes. **Only 3 students said that the atoms "separate",** as in the following extracts. This answer suggests a process of dilution rather than a chemical reaction, because although they say that the atoms separate, they don't comment that they regroup in a different arrangement, forming another substance.

"What happens is that the water splits them and the tablet becomes open, and the **molecules give way, mixing with the water."**

"When the effervescent **tablet** is dissolved in water, the atoms and **molecules split during the reaction."**

According to Filho and Celestino (2010) students often confuse terms such as mixture and reaction, dilute and dissolve, as well as having difficulty differentiating between chemical and physical transformations. They also showed that sometimes the microscopic view of a solution is very vague or non-existent.

Here it is clear that this type of confusion persists, and students have difficulty understanding what happens at the submicroscopic level, defining a chemical reaction as if it were just a dilution.

DV students A, B and D replied that during the reaction the atoms break apart and the molecules separate from each other. Student C said that the molecules will mix with each other, which was surprising compared to the other students' answers. This result is due to the fact that, as Vygotsky proposes, DV students develop creativity, memory and attention better in order to compensate for the lack of sight. Soler (1999) describes that hearing stimulates DV learners to adopt attitudes of attention, i.e. they really listen and pay attention to what is explained to them. Therefore, it is possible that they have already learnt this content and are able to remember the explanation for the phenomena. Furthermore, this research indicates that they have more creativity to imagine the molecular world, making the abstraction that is so necessary for the submicroscopic explanation of chemical phenomena. Whereas sighted students are often limited to what they can see, and have difficulty imagining the microscopic part, as proven in the studies by Soler (1999).Everyday situations were proposed so that the students could rate whether or not they thought knowledge of chemistry was important in their lives in each case. Figure 11 shows the result of this analysis.

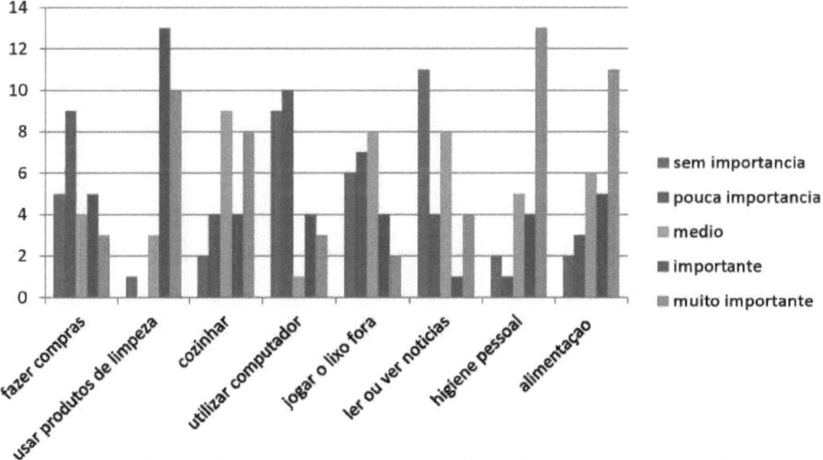

Figure 11 - Importance for students of chemical knowledge in everyday life

We can see in figure 11 that it was considered important to know chemistry when using cleaning products, in personal hygiene and in food, with the prevailing conceptions being that chemistry is something artificial, related to industrialised products. Shopping, using the computer and watching the news were the items least related to chemical knowledge. This shows that we are still in our infancy when it comes to scientific literacy, since technology is strongly linked to science, and the opinions that may appear in situations that appear in the newspapers are often related to this knowledge. For Chassot (2003), being scientifically literate means knowing how to read the language in which nature is written. With this knowledge, people will be able to control and predict natural transformations and everyday phenomena, and will be able to solve or avoid concrete problems in their lives.

The resources most used in chemistry lessons are texts, images, graphs and tables, formulae, equations and models. Blind students find formulae and equations difficult or very difficult to understand, as well as models (atomic, molecular, etc.) and find figures and images easy or very easy. The ADVs considered the texts and representations to be easy or average, while the models were difficult (all answered difficult). Graphs, tables and figures were considered difficult or very difficult, which is to be expected since they are very visual representations.

The vast majority of sighted students (19) could not recall any activity or resource in class that had helped them understand the content. The others mentioned experimental activities. The VI students couldn't think of any didactic activities or resources either.

What we noticed during the observation and filming of the class was that the ADVs took a little longer than the others to do the same activities. While they weren't finishing the task, the sighted students were idle and the classroom was in turmoil, jeopardising the application of the proposed tasks.

Another relevant observation in this study was that the ADV group is separated from the others, sitting in the first two desks of the middle rows, close to each other, but interacting little with the rest of the class. Among the ADVs, the student with low vision was the most participative during the explanations.

In a conversation recorded after the didactic activities, the students with disabilities said that there is a veiled exclusion, that no matter how much they participate and how receptive their sighted classmates and teachers are, they still end up in a group of their own, separated from their classmates. They believed that this was something that happened naturally, unintentionally.

Even so, they said that they felt good when they were doing activities in mixed groups of ADVs and psychics, as they were able to interact more and exchange experiences with their colleagues.

6.1.2 Practical lessons

The practical lessons were divided into three parts. In the first lesson, some experiments were carried out in which the students identified whether or not the phenomena were chemical reactions, describing the initial and final characteristics of the systems, in order to conclude what evidence accompanied the phenomena. In order to follow a multisensory methodology, as proposed by Soler (1999), it is important for the teacher to encourage the students to use all their senses to try to perceive the differences, with the necessary precautions.

The reactions carried out were dissolving an effervescent tablet in water, mixing sodium bicarbonate with vinegar and mixing hydrochloric acid with sodium hydroxide. In addition to the visual evidence, the students were expected to identify the sounds of gas being released, to use touch to feel the bubbles release and the change in temperature, the change in odour due to the consumption of vinegar, and also to observe that some reactions do not show macroscopic changes in the system.

In the experiment where there was evidence (mixing hydrochloric acid with sodium hydroxide), the students could easily recognise why there was a chemical reaction. The last experiment generated some doubt, with students questioning whether or not a chemical reaction was taking place. In order to confirm that a reaction was taking place, the mixture was redone using phenolphthalein as an indicator, so that the sighted students could see that there had been a change of substances, even when we couldn't see this through evidence, and they described to their blind classmates what was being visualised. This helped learning because, according to Piaget's socioconstructivist theory (1977), when there is an imbalance in the cognitive system, i.e. the student's previous knowledge cannot explain the phenomenon being observed, they need to assimilate and accommodate the new facts, creating a new state of equilibrium, modifying the previous mental schema, which had some incorrect ideas.

In the second practical lesson, which was twinned with the first, some more phenomena were observed, with the aim of verifying that some physical phenomena can have macroscopic changes in the system, without forming new materials, i.e. the evidence does not guarantee that a chemical reaction has taken place.

The experiments carried out were dissolving urea in water, dissolving sodium hydroxide and mixing mints with soda. In these cases the students observed the changes in the system, always emphasising the importance of using all the senses.

Some students still believed that the dissolutions of sodium hydroxide and urea could be reactions due to the change in temperature, but in the case of the soft drink they all identified that there was no reaction because the gas wasn't formed, but was already there.

This was followed by an explanation and discussion to reinforce that a chemical reaction represents the recombination and formation of new bonds between the chemical elements of the reactants to form products. This does not imply visible changes or irreversibility as some textbooks present, because according to Mortimer and Machado (2011), chemical reactions are generally accompanied by physical transformations, which make it possible to recognise their occurrence. What we can recognise are the physical transformations, because there is no direct evidence that the phenomenon that has occurred is a chemical reaction. It is our accumulated empirical knowledge that allows us to identify, through these physical transformations, the cases in which new materials are produced and, therefore, chemical reactions.

The last experimental lesson took place the following week, in two lessons of 50 minutes each. Three experiments were carried out: mixing sodium hydroxide and copper sulphate II, sodium bicarbonate with vinegar (acetic acid) and burning a steel sponge, always measuring the mass of the system before and after the reaction. The aim was to experimentally observe what happens to mass in chemical reactions in an open system, and then discuss the law of conservation of mass.

The ADVs used two-pan scales to manually visualise whether there was an imbalance in the system, thus observing the change in mass. The sighted students used a digital scale.

For the first reaction, the students observed that the mass remained the same, unchanged, before and after the reaction.

In the second case, the mass decreased due to the loss of carbon dioxide to the environment from the effervescence. One of the groups explained that the effervescent tablet **"disappeared" into the water. Four groups,** including the ADV group, said that the gas left the system, and two groups didn't explain.

In the last experiment, the mass increased due to the reaction of the iron with the oxygen in the air. Only one group of sighted students and the ADV group observed the increase in mass, the other groups ended up over-handling the sponge, losing some material. No group was able to explain why the mass increased.

To conclude this lesson, a collective discussion was held, returning to the law of conservation of mass, leading the students to conclude that the mass does not change because the atoms involved in the reaction are the same before and after. Emphasis was placed on the fact that if the system were closed, the mass would remain the same in reactions 2 and 3, but as the reactions involved gases, there was a decrease or increase in mass.

The class in which the research was carried out was, according to the teachers and the observations in the filming, extremely agitated. According to the school's pedagogical team, a group with more learning difficulties was formed to stay with the ADVs, so that the pace of the lessons would be reduced compared to the other classes.

There are a few points worth highlighting here, such as the fact that at the beginning some of the students were resistant to DV students carrying out experiments, believing that they wouldn't be able to do or observe anything. This view was soon overcome in the first experiment, as they realised that there were several ways to observe a chemical reaction.

It was interesting to note that in the classes where the group of visually impaired students was split up, i.e. each of them in a group with sighted colleagues, the class was more productive, there was less commotion, and they were helping and trying to observe the phenomena from the perspective of the blind. On the other hand, in the class where there was a group of visually impaired students and other groups with sighted students, the commotion was much greater and the activity of some groups was not carried out correctly.

Soler (1999) states that joint activities, in which the sighted help and describe phenomena to the DV, reduce the effects of distraction on students who have problems concentrating. This can be clearly observed at this stage of the research.

After the first sequence of experiments, the sighted students gave a written account of how they felt the interaction with their DV colleagues went. All the groups reported that the participation and interaction was good and that the VDAs helped them to perceive non-visual evidence, helping them to understand the phenomena.

Therefore, inclusion in some cases can be good not only for students with special needs, but for the group as a whole. With a more heterogeneous group, the social environment is diversified. In this case, this was a positive factor, as mediation was created in which the easier students helped those who had difficulties and the blind students showed the sighted how they could see that reaction in a different way. This can be explained by Vygotsky's (1991) theory of the Zone of Proximal Development, as it was observed that the students developed their intellect within the intellect of those around them, in other words, to be able to develop the expected knowledge they needed the teacher's support, and the students who were able to appropriate it helped their colleagues.

The final questionnaires were made up of open questions or Likert scales and were answered by 21 students and 4 ADVs. With regard to the experiments, most of the sighted students (16) said that they enjoyed carrying out the experiments, and 15 of them fully or partially agreed that they were able to relate them to the theories and equations seen in the classroom. Only 6 students said they hadn't been able to make this connection and 2 didn't answer. Student A partially agreed that she was able to relate the experimental practices to the theory, while students B, C and D fully agreed, and all said they had enjoyed doing the experiments.

6.1.3 Working with Braille chemical spelling

In the next lesson, a week later, the text **"Chemical reactions and the evidence"** was read, **followed by a recap of what had been** worked on in the experiments. Next, we explained how we represented the phenomena through equations, using the same reactions carried out in the laboratory in the previous lesson, working on Braille chemical spelling with the blind students so that they could follow the equations along with the sighted students.

The DV students didn't know Braille chemical spelling, but they had no trouble using it. Student A commented that

> "Because I didn't know Braille yet, I was only able to understand the formulae after explaining how they were put together."

Therefore, in order to be able to use Braille chemical spelling, it is necessary for the teacher to know and understand it, because without guidance the students would not be able to understand some details. The main difficulty was the fact that there was no number sign before the lower indices of the formulae, and the students commented that they could confuse them with letters, since in Braille the numeral 1 has the same sign as the letter A, the numeral 2 as the letter B, and so on.

Even so, after a little practice, they showed that they understood and were able to read the equations themselves, they understood what reactants and products were, the physical states of the substances, thus relating the theory to what they had previously seen in practice.

6.1.4 Use of the molecular model

This activity began with a review of the experiments from the previous lesson and a discussion about why mass is conserved in chemical reactions.

It was explained to the students that in a chemical reaction, the reactants are consumed and new products are formed, through the breaking of chemical bonds and the reorganisation of the participating atoms. This means that when a reaction starts, the bonds in the molecules of the reactants break, the atoms separate and reorganise themselves to form the products. Therefore, the number of atoms present in the product of the reaction must be equal to the total number of atoms in the reactants, so the mass will not change.

Based on this discussion, the balancing of chemical reactions was demonstrated, working with the molecular model. The first example was done together, so that the students could observe and understand how they would set up the representation of the atoms in the chemical reaction.

Figure 12 shows how the decomposition reaction of hydrogen peroxide looks before balancing, having assembled it molecule by molecule with the students, noting the number of atoms of each chemical element in the formulae. The Braille symbols are on the pieces of paper glued to each sphere.

$$H_2O_2 \ (l) \rightarrow H_2O \ (l) + O_2 \ (g)$$

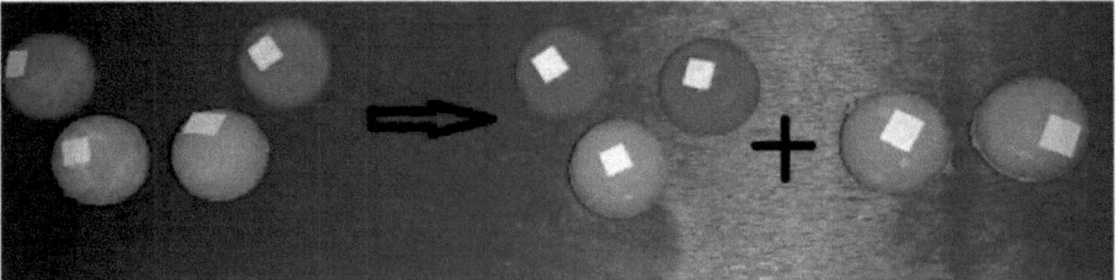

Figure 12 - Decomposition of unbalanced hydrogen peroxide.

Figure 13 shows the reaction after balancing, done by counting the number of atoms of each chemical element and adding molecules until they were equal on both sides of the equation.

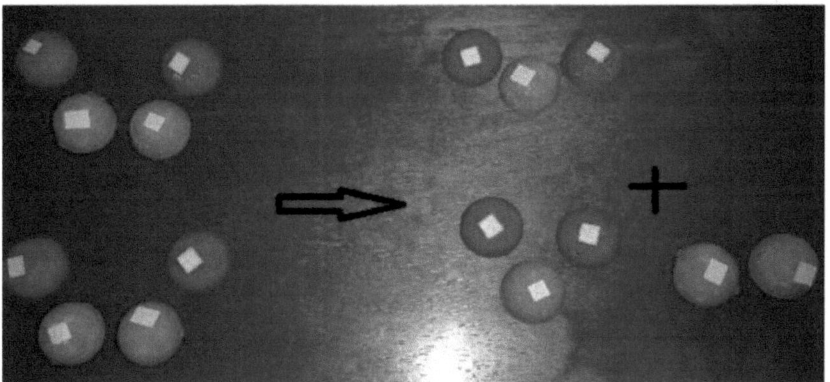

Figure 13 - Decomposition of hydrogen peroxide after balancing.

The sighted students were able to work well with the model, setting up the reactions and arriving at the correct stoichiometric coefficients.

Most of the students fully (8) or partially (3) agreed that the molecular model was easy to understand, and only 2 totally (1) or partially (1) disagreed. The rest (6) had a neutral opinion.

As for helping to understand what happens in chemical reactions with atoms, the majority also agreed (13) that it helps, while only 3 disagreed.

Most of the students said they enjoyed working with the molecular model and thought it helped when balancing the reactions. It was only with regard to balancing the reaction without using the model that opinions were divided. Six (6) students fully or partially agreed that they could do the balancing even without the model, 4 had a neutral opinion and 9 disagreed, assuming that they would need the molecular model to do the balancing. Therefore, the students did not feel prepared to balance the reactions without needing the model.

The ADVs, on the other hand, had some difficulties at this stage, thinking that the letters were difficult to identify, because they picked up the ball in the wrong position, with the symbol facing downwards, for example, and they couldn't identify the atom.

They didn't find the model easy to understand, but they still partially agreed that it helped them understand what happens to atoms and molecules in chemical reactions. As for balancing, the blind students disagreed that they could do it with the molecular model, since they had difficulties working with it. They also disagreed that they could do the balancing without using the model, because they couldn't understand this concept. Only the student with low vision said he was able to understand balancing.

In an interview after the end of the activities, the DV students said that the idea of the balls with magnets was a good one, as it prevented them from rolling around and dropping them. They suggested that each chemical element be represented with balls of different sizes or textures, to make it easier to perceive.

At the start of the project, the students defined chemical reactions as artificial and physical transformations as natural. At the end of the work carried out with the sequence of activities, only 2 students maintained this concept and 1 student used the reversible vs. irreversible criterion. The rest (22), including the DVs, replied that chemical reactions are not observable to the naked eye, that new substances are formed and that physical transformations are what we can visualise, without necessarily changing the nature of the material. According to Mortimer and Machado (2011), chemical reactions are usually accompanied by physical transformations, which make it possible to recognise their occurrence. What we can recognise are the physical transformations, because there is no direct evidence that the phenomenon that has occurred is a chemical reaction. It is our accumulated empirical knowledge that allows us to identify, through these physical transformations, the cases in which new materials are produced and, therefore, chemical reactions. It was interesting to note that before the activities the students did not have a clear distinction between what was a chemical reaction and what was just a physical phenomenon, and very few (2) were able to correctly explain what happened to the atoms and molecules during a reaction. In the final questionnaire, the result was positive, as 15 sighted students and 4 ADVs said that a chemical reaction occurs when atoms meet and give rise to new substances, 3 related it to physical changes and only 2 cited the criterion of irreversibility. In this way, the application of the didactic sequence was effective for everyone, as most of them changed the incorrect conceptions they had at the beginning. Figure 14 shows the concepts related to chemical reactions on the part of the sighted students.

Figure 14 - Students' definition of a chemical reaction.

This was confirmed in the following questions, in which they identified whether tearing or burning a sheet of paper was a chemical reaction. The result was that 20 students, including 3 of the DVs, said that tearing the paper was not a reaction, and 17 said that burning the paper was a reaction. Only student C switched concepts. Once again, the few students who didn't answer as expected still related the chemical reaction to macroscopic factors or irreversibility.

To back up this analysis, the students were asked what happens when we add salt to water. 18 of them, including 3 of the DVs, wrote that the salt only dissolves, and that this was a physical transformation. Only 4 of them and student A believed that there was a chemical reaction.

Therefore, some students still had the obstacle of first experience, which, according to Bachelard (1971), is a kind of pre-scientific generalisation that ends up making knowledge very vague. Saying that a physical phenomenon was reversible and a chemical phenomenon irreversible is a concept that was part of textbooks until recently, so some students may have fixed this concept, creating this type of epistemological obstacle.

But in general, the objective was achieved, as most of the students transformed or rectified their knowledge of reactions, starting to see them firstly as breaking bonds and reorganising atoms into molecules different from the initial ones, and then observing the macroscopic consequences of this, i.e. the physical evidence that can be seen.

In addition, other positive points were observed, such as the real inclusion of ADVs during lessons in group experiments and carrying out the software activities at the same time and in the same way, more effective teamwork and the development of collaboration between students.

6.1.5 Application of the computer programme

In the last lesson of the sequence, on the same day as the molecular model, the Inclusive Chemistry programme was applied. The students showed interest and satisfaction. They responded well to the activities and needed help with some topics, but in general they managed to navigate without difficulty, showing a good knowledge of the content.

To check the efficiency of this product, two evaluation tools were applied, one proposed by Relvas (2005), with some adaptations, and Nokelainen's questionnaire (2006), which evaluates criteria such as student control, student activity, goal orientation, applicability, motivation, evaluation of prior knowledge and flexibility of the software.

The DV students said they enjoyed using the programme and that it helped them learn the content. Only student A had a neutral opinion.

All the students, DV or not, agreed that when they used the programme they had to think and make their own decisions in order to arrive at the solution, and that they, and not the programme, had control over the responsibility for their learning.

Regarding the fact that the material is divided into sections so that students learn it in a predefined order, all the VI students partially disagreed, while 17 of the sighted students totally disagreed and 4 partially disagreed. Possibly this result is due to the fact that they could move on from one question to another when they were having difficulties, not necessarily having to follow an order, which is a positive point for the proposal of the digital teaching material developed in this master's programme.

Students B and C thought that the programme held their attention, making them forget what was going on around them, while students A and D didn't think so. Opinions were also divided among the other students, with 10 partially agreeing, 6 being undecided, and 5 partially disagreeing. Therefore, a future improvement to the software would be to make it a little more attractive to the sighted, with animations and questions in more elaborate formats.

By using the programme, the ADVs felt proud of their problem solving, realised how much progress they had made in their studies and that they knew more about some topics than others. The result was similar among the sighted. Therefore, this product allows the student to make a self-evaluation, observing their difficulties and limitations, thus allowing them to strive to seek knowledge that has not yet been fully consolidated.

With regard to applicability, 3 of the DV students said that the programme teaches skills that they will need in the future and that they are able to use them, including in exams. One student disagreed that the programme teaches skills that they will need in the future, stating only that the skills would be useful when taking tests. Of the rest of the class, 12 students totally or partially agreed that the programme teaches skills they will need, 5 were undecided, and 4 partially disagreed. As for the part about helping to carry out assessments, all the sighted students agreed totally or partially.

Similarly, the same 3 DV students indicated that the material was based on the idea that one learns best by doing for oneself, as well as that it was adequately challenging, while 1 partially disagreed with these statements. The students with normal vision partially (6) or totally (12) agreed with the first statement, and 3 partially disagreed. As for being challenging in the right way, 15 agreed and 6 disagreed. Observing the students as they carried out the activities in the software, some really struggled with some questions and may have found the material too challenging, as the class had learning difficulties. Even so, the result was good, since most of them found the challenges interesting and managed to answer them, overcoming their difficulties.

In terms of motivation, the software proved to be satisfactory, as all the VI and VI students agreed that they tried to achieve a high score when solving the activities proposed in the software and that they were interested in learning what they were doing.

Prior knowledge was valued, as all the students agreed that the material required knowledge they had previously seen in other materials, and that they could use their knowledge acquired in class or in their everyday lives.

The ADVs felt that they had to remember many things at once when using the programme, with the exception of student A. Of the sighted students, 10 partially agreed, 4 were undecided and 7 partially disagreed. This division indicates that the programme covers all the content that has been worked on in class, encouraging students to remember all of it when carrying out the tasks.

Student A disagreed that the educational object presented new material or recapitulated old material in suitable portions for her, while the other 3 students agreed. But all of them said that they learnt more quickly with this material than they normally do. The students without sight problems agreed with the statement, with only 3 disagreeing that they learn better with the programme than without it.

Regarding the understanding of the[1] interface of the programme, the sighted showed good acceptance, as shown in figure 15 below.

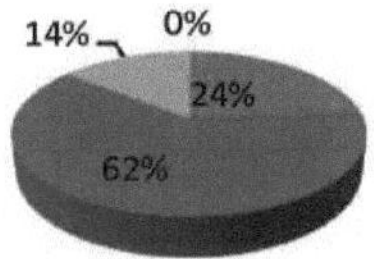

Figure 15 - Understanding the software interface

The[10] interface was therefore judged to be excellent and easy to understand. All the students gave it an excellent rating for the visualisation of the windows and the fonts used.

The evaluation of the logical sequence of the information is shown in the graph in figure 16 below.

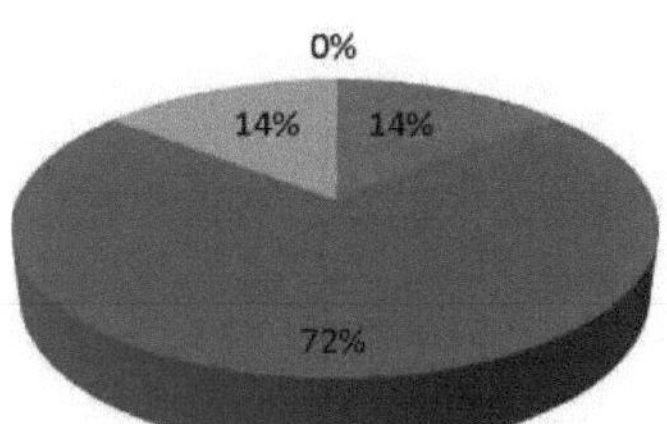

Figure 16 - Information is organised logically

Some students had some difficulties, but in general the organisation of the information was considered to be excellent or good by the majority.

The question of whether the didactics used were in line with the student's knowledge of the subject was considered excellent by 13 of the interviewees and good by 8 of them. The way in which the elements are presented and their consistency with the students' knowledge was given as excellent by 7 people and good by 14. Therefore, the content worked on in the didactic sequence can be considered to be in line with what appears in the programme.Figure 17 shows the opinion of the sighted students

---

[10] An interface is considered to be the means by which a program communicates with the user, including a command line, menus, dialogue boxes, an online help system, etc.

on the ease and clarity of the solutions when using the software.

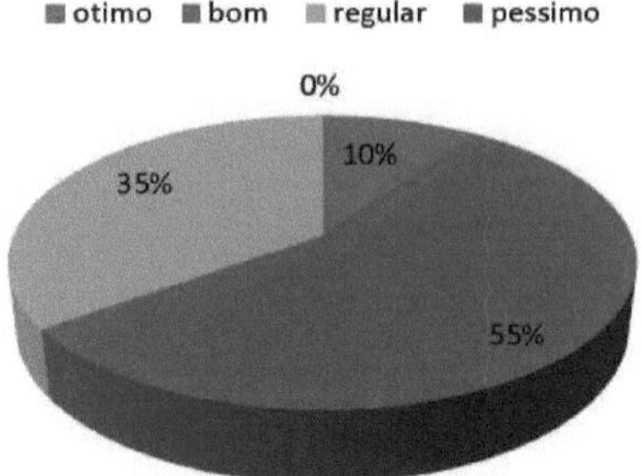

Figure 17 - Visualisation of solutions

The clarity and understanding of the solutions to the questions entered into the software were considered good by most of the students, but a few thought they were excellent, and some thought they were fair. During the application, it was noted that some students had difficulty typing their answers in the way the programme requested, which may be the reason for this result. Another future improvement could be to add a correction for incorrect answers, i.e. when you type in the wrong answer, the programme opens a window explaining why that can't be the answer.

As for holding the student's attention, the software is good, with 16 students evaluating it this way and 5 giving it an "excellent" rating.

All the ADVs found the interface friendly and easy to understand, the "visualisation" of the windows easy, the audio clear and easy to understand, the information logically organised, the didactics used and the way the elements were presented in line with their level of knowledge, and that the software held their attention.

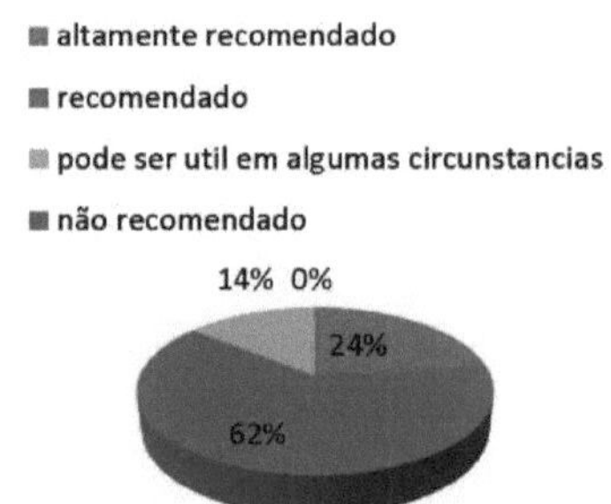

Figure 18 - Result of the software evaluation

Students B and C considered the software highly recommended, student D considered it recommended and student A reported that it could be useful in some circumstances.

In an interview after the activity, the students said they really liked the initiative to create

54

accessible software for them, saying that they knew very few in all areas of knowledge. In chemistry, they said they had once worked with a programme on the periodic table, but that they didn't find the results so good because the table for the blind doesn't have the same format and layout as the normal table, so that when the teacher explains, he talks about the normal table, and they end up not understanding or being able to use the one that has adaptations for the blind.

This implies that teachers really need initial and continuing training to work effectively with this audience, because if they don't know the braille chart, they don't know how it's laid out, so how can they explain it to the students?

Therefore, the evaluation of the software was satisfactory, as it achieved most of the planned objectives, requiring only a few modifications.

In Brazil, there are very few references to the use of software in teaching the visually impaired. At an international level, Brown, Pettifer and Stevens (2003) have achieved good results by developing a programme that reads Kekulé structures, describing each molecule and bond in detail and providing a wealth of information for the blind. The website http://www.molinsight.net/ also brings together some molecule editor software, including navmol, which has already been created and tested, for blind people to download and use.

The conclusion is that the sequence of activities developed, as well as the materials and methodologies created, led students with and without visual impairments to learn the content of chemical reactions in an equal, participatory and inclusive way.

The manual and materials can be used in any school, as they are low-cost proposals using common materials, and the experiments can even be carried out in the classroom if the school is not equipped with a science laboratory. The software can also be used in any school that has a computer lab. The results therefore indicate that the proposals are viable for any teacher who wants to use them in their lessons, whether or not they have ADV in their class.

The software was evaluated as good, constituting a breakthrough in the area of inclusion, given that there are practically no digital chemistry objects accessible to the blind.

The intention of promoting the real inclusion of ADVs during lessons was realised, showing that we must first have the will to do it and the initiative to make it happen. Observing the interaction and exchange of experiences between blind and non-blind people during the implementation of the project shows that we succeeded in our primary objective in this dissertation.

CHAPTER 7

FUTURE WORK

With the application of this project, there is the prospect of improving the molecular model in the future, making it more useful for ADVs, using the suggestions given such as texturisation and size differentiation.

The software was developed quickly, simply and objectively, and is very effective for blind people. In the future, the visual aspects could be improved so that blind students are more interested in the material. It would also be interesting to add feedback on incorrect answers in the form of explanations, as this would favour the learning of content that had not been clear up to the point of using the software.

REFERENCES

BRAZIL. Decree No. 5.296 of 2 December 2004. Available at: http://www.planalto.gov.br/ccivil_03/_ato2004-2006/2004/decreto/d5296.htm. Accessed on: 20/12/12.

BRAZIL, National Curriculum Guidelines for Basic Education. Brasília: Ministry of Education, 2013.

BRAZIL, LDB. Law 9394/96 - *National Education Guidelines and Bases Law.*

BRASIL. *National Policy for Special Education from the Perspective of Inclusion.* Brasília: Ministry of Education; 2008. Available at: www.portal.mec.gov.br

BROWN, A., PETTIFER, S. STEVENS, R. *Evaluation of a NonVisual Molecule Browser.* Proceedings of the 6th international ACM SIGACCESS conference on Computers and accessibility, Atlanta, GA, USA, 40 - 47. 2003.

BACHELARD, Gaston (1971). Epistemology. Lisbon, Edições 70, 2006.

BERTALLI, J. G. *Teaching molecular geometry to visually impaired and non-visually impaired students using an alternative atomic model.* 2010. Dissertation - UFMS. Campo Grande - MS.

BEYER, H.O. *A educação inclusiva: ressignificando conceitos e práticas* da *educação especial,* Inclusão: Revista da Educação Especial / Secretaria de Educação Especial. V. 2, n. 2, Brasília, 2006.

BOGDAN, R. BIKLEN, S.K. *Qualitative research in education: an introduction to theory and methods.* Portugal: Porto Editora, 1994-2006. 336 p. (Ciência da educação ; 12)

CANZAN, R. MAXIMIANO, F.A. *Changes in chemical equilibrium systems: analysis of the main illustrations in textbooks.* In: XV Encontro Nacional de Ensino de Química (Xv ENEQ), 2010, Brasília.

CARVALHO, R. E. *Removing barriers to learning. In: Leap into the future: current trends /* Secretariat for Distance Education. Brasília: Ministry of Education, SEED, 1999.

CARVALHO, A. M. P. *O uso do vídeo na tomada de dados: pesquisando o desenvolvimento do ensino em sala de aula. Pro-Posições,* v.19, n.1, p.5-13, 1996.

CARVALHO, J. O. F., BEZERRA, C. M. C., BUTTON, V. L. D. S. N., DALTRINI, B. M., KAWAI, A. K., BARROS, R. *BRILLE: An interface system to facilitate communication between the visually impaired student and the sighted teacher.In:* II Seminário ATIID - Acessibilidade, TI e Inclusão Digital, São Paulo-SP, 2003.

CHAGAS, J. A. S. das. *Obstacles encountered in the process of understanding the concept of chemical reaction.* Dissertation (Masters in Education). UFPE. 2007.

CHASSOT, Attico. Scientific literacy: a possibility for social inclusion. Revista Brasileira de Educação, Rio de Janeiro, 22: 89-100, 2003.

CYSNEIROS, Paulo G. New Technologies in the Classroom: Improving Teaching or Conservative Innovation? In IX ENDIPE, Anais II, vol. 1/1, pp. 199216. SP, Águas de Lindóia, 1998.

DIAS, C. O. De Olho na Tela: Accessibility Requirements in Learning Objects for Blind and Visually Impaired Students. 2010. 162 f. Dissertation (Master's in Education) - Postgraduate Programme in Education, Federal University of Rio Grande do Sul, Porto Alegre, 2010.

DUARTE, R. Pesquisa qualitativa: reflexões sobre trabalho de campo. Cadernos de Pesquisa, Campinas, n. 115, p. 139-154, Jul. 2001.

ENGEL, G. I. Research-action. Revista Educar, Curitiba, n. 16, p. 181-191.
Editora daUFPR. nov. [2000]. Available at:
<http://www.educaremrevista.ufpr.br/arquivos_16/irineu_engel.pdf>. Accessed on: 28/08/2013.

FILHO, J.R.F. CELESTINO, R.M.C.S. Investigation of the construction of the concept of chemical reaction based on previous knowledge and social interactions. Ciências & Cognição, vol. 15 (1), p. 187-198, 2010.

GARCÍA RUIZ, M. CALIXTO R. Las Actividades Experimentales como una *Estratégia de Ensenanza de la Ciencias Naturales en la Educación Básica.* Perfiles Educativos, n. 83-84, p. 105-118, 1999.

GIORDAN, M. *The role of experimentation in science teaching.* Química Nova na Escola. São Paulo. n.10, p.43-49. 1999.

GIORDAN, M. *O computador na educação em ciências: breve revisão crítica acerca de algumas formas de utilização.* Ciência e Educação, v.11, n.2, p.279- 304, 2005

GONÇALVES, F. P., REGIANI, A. M., AURAS, S. R., SILVEIRA, T. S., COELHO, J. C., & HOBMEIR, A. K. T. *Inclusive Education in Teacher Training and Chemistry Teaching: Visual Impairment in Debate.* Química Nova na Escola. São Paulo. V. 35, n.4, p.264-271. 2013.

HONTANGAS, N.A. Puente, J.L.B. *Atención a la diversidad y desarrollo de procesos educativos inclusivos.* Prisma Social: revista de ciencias sociales, Madrid, n.4, jun. 2010.

JOHNSTONE, A.H. *The Development of chemistry teaching: A changing response to changing demand.* Journal of Chemical Education n. 70, 701-704. 1993.

JUSTI, R. S. & RUAS, R. M. *Chemistry learning: reproduction of isolated pieces of knowledge?* Química Nova na Escola, n. 5, p. 24-27, May 1997.

KOERICH, M. S; BACKES, D. S; SOUZA, F. G. M; ERDMAN, A. L; ALBURQUERQUE, G. L. *Research-action: a methodological tool for qualitative research.* Revista Eletrónica de Enfermagem. V. 11, n. 3, p. 717723. 2009.

LEVY, P. *Cibercultura*. Rio de Janeiro: Ed. 34. 1999.

LOPES, A. PASSERINO, L. M. ROSA, M.V. EMILIANO, M. V. *The development of Learning Objects for visually impaired students based on accessibility requirements*. In: VI Conferencia latinoamericana de objetos de aprendizaje y tecnologias para la educación laclo. 2011.

MARTINS, I. *Data as dialogue: building data from observation records of discursive interactions in science classrooms*. In: SANTOS, F.M.T. & GRECA, I.M. (eds.). Science teaching research in Brazil and its methodologies. Ijuí: Editora Unijuí, 2006.

MINER, D. L., NIEMAN, R. SWANSON, A. B., WOODS, M. E. *Teaching Chemistry to Students with Disabilities: A Manual for High Schools, Colleges, and Graduate Programs*, 4th ed.; American Chemical Society Committee on Chemists with Disabilities: Washington, D.C., 2001.

MINISTRY OF EDUCATION. *Braille Chemical Spelling - For use in Brazil*. Department of Special Education - Brasília: Ed. 2. MEC; SEESP, 2011.

MÓL, G.S. RAPOSO, P.N. SANTOS, G.A. NETO, J.D. BRITO, A.G. *The inclusion of visually impaired students as a theme in dissertations and theses in Postgraduate Programmes in the Science and Mathematics Teaching Area of Capes*. In: XV Encontro Nacional de Ensino de Química (XV ENEQ), 2010, Brasília.

MORTIMER, E. F.; MACHADO, A. H.; ROMANELLI, L. I. *A proposta curricular de química do Estado de Minas Gerais: fundamentos e pressupostos*. Química Nova, v. 23, n. 2, p. 273-283, 2000.

MORTIMER, E. F. MIRANDA, L. C. *Transformations: Students' Conceptions of Chemical Reactions*. Química Nova na Escola, n. 2, p. 23-26, nov.1995.

NASCIMENTO, C.C. COSTA, S.S.L. AMIN, L.H. *Rethinking chemistry teaching: A proposal for the visually impaired.* In: COLÓQUIO INTERNACIONAL EDUCAÇÃO E CONTEMPORANEIDADE, 4., 2010, Laranjeiras. *Proceedings...* Laranjeiras, 2010.

NOKELAINEN, Petri. *An emprical assessment of pedagogical usability criteria for digital learning material with elementary school students*. Educational Technology & Society, v.9 (2), p. 178 - 197, 2006.

NUNES, B. C.; DUARTE, C. B.; PADIM, D. F.; MELO, I. C.; ALMEIDA, J. L.; TEIXEIRA JÚNIOR, J. G. *Proposals for experimental activities designed by future chemistry teachers for visually impaired students*. In: Proceedings of the XV National Meeting on Chemistry Teaching, Brasília: 2010.

OMOTE, S. Inclusion and the question of differences in education. Perspectiva, Florianópolis, v. 24, p. 251, 2006.

PATTON, M Qualitative evaluation methods. Beverly Hills, Sage Publ., 1986.

PEREIRA, F. SOUSA, J. A. MATA, P. LOBO, A. M. Development in the teaching of Chemistry to blind and partially sighted people. Boletim da sociedade portuguesa de Química, n.112, 7-15, 2009.

PIAGET, Jean (1977c): The Development of Thought - Equilibration of cognitive structures. D Quixote Publications. Lisbon.

PIRES, R.F.M. Proposal for a guide to support the pedagogical practice of chemistry teachers in inclusive classrooms with students who are *visually impaired.* 2010. Dissertation - UnB. Brasília - DF.

REITZ, Doris Simone. *Evaluating the Impact of Technical and Pedagogical Usability on Learner Performance in E-learning. Thesis for Doctorate in Informatics in Education.* Interdisciplinary Centre for New Technologies in Education - Federal University of Rio Grande do Sul. p. 197. 2009.

RELVAS, E. An *Evaluation Tool for a Chemistry Educational Software Product.* 2005.

RESENDE FILHO, J. B. M. et al. *Evaluation of the Level of Knowledge of High School Students in the city of João Pessoa with Visual Impairment about Braille Chemistry and Mathematics Graphics.* Revista Educação Especial, v. 26, n. 46, p. 367-384, 2013.

REY, F.G. *Qualitative research in psychology: paths and challenges.* São Paulo: Pioneira Thomson Learning, 2002. 188p.

ROSS, P. R. *Learning and knowledge: foundations for inclusive practices.* Revista pERSPECTIVA, v. 24, n. Especial, p. 273-299, 2006.

SILVA, M.A.; CÉSAR, M. *Curricular adaptations in physical-chemical sciences: A path towards a more inclusive education.* In: Congresso Galaico- Português de Psicopedagogia, 8., 2005. Braga: University of Minho, 2005 (p. 605).

SILVA, E.L.; MENEZES, E.M. *Metodologia da Pesquisa e Elaboração de Dissertação.* 4ª ed. Florianópolis: UFSC Distance Learning Laboratory, 2005

SOLER, Miquel-Albert. *Multisensory science teaching: Un nuevo método para aluminos ciegos, deficientes vises, y también sin problemas de visión.* Barcelona: Paidós, 1999.

SUPALO, C. KREUTER, R. A. MUSSER, A. HAN, J. BRIODY, E. MCARTOR, C. GREGORY, K. MALLOUK, T. E. *Seeing Chemistry through Sound: A*

Submersible Audible Light Sensor for Observing Reactions in Real Time. Assistive Technology Outcomes and Benefits 2006, 3 (1), 110-116.

UNESCO. World Declaration on Education for All: plan of action to fulfil basic learning needs. Jomtien -
Thailand, 1990. Available at :
<http://unesdoc.unesco.org/images/0008/000862/086291por.pdf>

UNESCO. Salamanca Declaration and line of action on *special educational* needs. Salamanca - Spain, 1994. Available at: <http://portal.mec.gov.br/seesp/arquivos/pdf/salamanca.pdf>

VILELA-RIBEIRO, E. B. and BENITE, A. M. C. *Inclusive education in the perception of chemistry teachers.* Ciência & Educação, v. 16, n. 3, p. 585-594, 2010.

VYGOTSKI, L. S. *Obras Escogidas V: Fundamentos de defectología.* Madrid: Visor, 1997.

VYGOTSKI, L.S. *Thought and Language.* 3ª ed. São Paulo, Martins Fontes, 1991ª

ANNEX A - Nokelainen Questionnaire

Source: Nokelainen (2006) apud Reitz 2009

Only questions pertinent to the type of software developed were used, and the alternatives were totally agree, partially agree, undecided, partially disagree and totally disagree.

1. When I work on this task I feel that I, not the programme, have control over the responsibility for my learning.
    Criterion: STUDENT CONTROL

2. I have to think and make my own resolutions to learn this learning material.
    Criterion: STUDENT ACTIVITY

3. This learning material has been divided into sections, my task is to learn them in a predefined order.
    Criterion: STUDENT ACTIVITY

4. This learning material provides learning questions without a predefined model for solving them.
    Criterion: STUDENT ACTIVITY

5. I got so deep into this learning material that I forgot everything that was going on around me and how much time had passed.
    Criterion: STUDENT ACTIVITY

6. When I work with this learning material I feel that I know more about some topics than others.
    Criterion: STUDENT ACTIVITY

7. I'm proud of my solutions to the problem presented in the learning material.
    Criterion: STUDENT ACTIVITY

8. This learning material shows how much progress I've made in my studies.
    Criterion: GOAL ORIENTATION

9. This learning material is strictly limited.
    Criterion: GOAL ORIENTATION

10. This learning material teaches skills that I will need. Criteria: APPLICABILITY

11. I feel that I am able to use the skills and knowledge that this learning material has taught me in the future.
    Criterion: APPLICABILITY

12. This learning material is based on the idea that "one learns best by doing for oneself".
    Criterion: APPLICABILITY

13. I feel that this learning material will help me to perform better in the exams.
    Criterion: APPLICABILITY

14. This learning material is suitably challenging for me. Criterion: APPLICABILITY

15. I try to achieve a high rating as much as I can in this learning material.
    Criterion: MOTIVATION

16. I'm interested in the topics in this learning material.

Criterion: MOTIVATION

17. This learning material requires me to know something that has been thought of in some other learning material.
Criterion: EVALUATION OF PRIOR KNOWLEDGE

18. I can use my previous knowledge when I study with this material.
Criteria: STUDENT CONTROL, APPLICABILITY, EVALUATION OF PRIOR KNOWLEDGE

19. This learning material offers optional paths for my progress.
Criterion: FLEXIBILITY

20. When I use this learning material, I feel like I have to remember a lot of things at once.
Criterion: STUDENT CONTROL

21. This learning material presents new material (or recapitulates old material) in "chunks" that suit me.
Criteria: STUDENT CONTROL, APPLICABILITY

22. I think I learn more quickly with this material than I normally do.
Criterion: APPLICABILITY

ANNEX B - Software evaluation questionnaire

Source: Relvas (2005)

Evaluation Description

Excellent: Fully meets the requirements and has no regularities (above 90 per cent).

Good: Achieves satisfactory levels of the requirements and shows some regularity that does not compromise development (between 60 and 90 per cent).

Regular: Partially fulfils the requirements and has some regularities (between 20%| and 60%).

Poor: Does not meet the requirements and has total or partial irregularities (below 20 per cent).

Question

Does the educational software offer a user-friendly interface, is it easy to understand?
( ) None ( ) Bad   ( ) Good   ( ) Good

Are the windows of the Educational Software easy to
visualise?

( ) None ( ) Bad   ( ) Good   ( ) Good

Are the fonts used legible?

( ) None ( ) Bad   ( ) Good   ( ) Good

Is the information organised logically?

( ) None   ( ) Bad   ( ) Good   ( ) Good

Is the didactics used in line with your level of knowledge
of the subject?

( ) None   ( ) Bad   ( ) Good   ( ) Good

Is the way the elements are presented in line with your
knowledge?

( ) None   ( ) Bad   ( ) Good   ( ) Good

Is the visualisation of the solutions clear, i.e. easy to
understand?

( ) None   ( ) Bad   ( ) Good   ( ) Good

Did the educational software capture your attention?

( ) None ( ) Bad   ( ) Good   ( ) Great

Based on your answers, classify Educational Software according to
the options below;

Results

( ) Highly recommended.

( ) Recommended.

( ) It can be useful in some circumstances.

( ) Not recommended.

yes

I want morebooks!

Buy your books fast and straightforward online - at one of world's fastest growing online book stores! Environmentally sound due to Print-on-Demand technologies.

Buy your books online at
www.morebooks.shop

Kaufen Sie Ihre Bücher schnell und unkompliziert online – auf einer der am schnellsten wachsenden Buchhandelsplattformen weltweit! Dank Print-On-Demand umwelt- und ressourcenschonend produziert.

Bücher schneller online kaufen
www.morebooks.shop

Printed by Books on Demand GmbH, Norderstedt / Germany